Hands-On Robotics with JavaScript

Build robotic projects using Johnny-Five and control
hardware with JavaScript and Raspberry Pi

Kassandra Perch

BIRMINGHAM - MUMBAI

Hands-On Robotics with JavaScript

Commissioning Editor: Gebin George
Acquisition Editor: Shrilekha Inani
Content Development Editor: Abhishek Jadhav
Technical Editor: Prachi Sawant
Copy Editor: Safis Editing
Project Coordinator: Jagdish Prabhu
Proofreader: Safis Editing
Indexer: Mariammal Chettiyar
Graphics: Tom Scaria
Production Coordinator: Deepika Naik

First published: August 2018

Production reference: 1300818

Published by Packt Publishing Ltd.
Livery Place
35 Livery Street
Birmingham
B3 2PB, UK.

ISBN 978-1-78934-205-5

www.packtpub.com

This book is dedicated to my loving partner, Kevin. Without your constant support and infinite patience, this book wouldn't exist.

– Kassandra Perch

`mapt.io`

Mapt is an online digital library that gives you full access to over 5,000 books and videos, as well as industry leading tools to help you plan your personal development and advance your career. For more information, please visit our website.

Why subscribe?

- Spend less time learning and more time coding with practical eBooks and Videos from over 4,000 industry professionals

- Improve your learning with Skill Plans built especially for you

- Get a free eBook or video every month

- Mapt is fully searchable

- Copy and paste, print, and bookmark content

PacktPub.com

Did you know that Packt offers eBook versions of every book published, with PDF and ePub files available? You can upgrade to the eBook version at `www.PacktPub.com` and as a print book customer, you are entitled to a discount on the eBook copy. Get in touch with us at `service@packtpub.com` for more details.

At `www.PacktPub.com`, you can also read a collection of free technical articles, sign up for a range of free newsletters, and receive exclusive discounts and offers on Packt books and eBooks.

Contributors

About the author

Kassandra Perch is an open web developer and supporter. She began as a frontend developer and moved to server-side with the advent of Node.js and was especially enthralled by the advance of the NodeBots community. She travels the world speaking at conferences about NodeBots and the fantastic community around them.

Thank you very much to my very patient editors at Packt—I brought things down to the wire more than once and you all have handled it exceptionally. To my mentor, Ray—you taught me that there's always room for improvement. To my mother, Kelly, father Joe, sister Kaitlynn, and brother Alex—you have been there for me my whole life, and I love you all dearly. Finally, to Raquel Vélez and Rick Waldron: your NodeBots workshop at JSConf 2013 changed my life, and you're both wonderful stewards of the community.

About the reviewers

Amit Rana is a Passionate Electronics Engineer, Maker, an Embedded Systems Professional, and Trainer. He has founded and is running three different firms in Electronics R & D, Product Development, and Robotics education. He holds a master's degree in electronics engineering. He has over 10 years of experience in embedded system designing and programming using various microcontrollers, Arduino, and Raspberry Pi with wireless technologies. He is also a professional writer who writes blogs on technology and education. He writes assignments on technical documents for few clients and also writes blogs on his website.

Shahid Memon is an analytical master of science in autonomous robotics engineering graduate possessing a bachelor's degree in computer science. He has collaborated with colleagues on product feasibility studies and new product ideas to meet clients' needs and support the company objectives. He has coordinated several product development projects and assisted in the design and testing phase. He is a strategic thinker with the ability to drive company goals and analyze research impacting products and business needs. He is an avid researcher of the latest trends within the technology industry and how it affects the business. He is a proven leader having outstanding communication, interpersonal, project management, and supervisory skills.

Packt is searching for authors like you

If you're interested in becoming an author for Packt, please visit `authors.packtpub.com` and apply today. We have worked with thousands of developers and tech professionals, just like you, to help them share their insight with the global tech community. You can make a general application, apply for a specific hot topic that we are recruiting an author for, or submit your own idea.

Table of Contents

Preface

There has been a rapid increase in the use of JavaScript in hardware and embedded device programming. JavaScript has an effective set of frameworks and libraries that support the robotics ecosystem.

Hands-On Robotics with JavaScript starts with setting up an environment to program robots in JavaScript. Then, you will dive into building basic-level projects such as a line-following robot. You will walk through a series of projects that will teach you about the Johnny-Five library, and develop your skills with each project. As you make your way through the chapters, you'll work on creating a blinking LED, before moving on to sensors and other more advanced concepts. You will then progress to building an advanced-level AI-enabled robot, connect their NodeBots to the internet, create a NodeBots Swarm, and explore MQTT.

By the end of this book, you will have gained hands-on experience in building robots using JavaScript

Who this book is for

Hands-On Robotics with JavaScript is for individuals who have prior experience with Raspberry Pi 3 and like to write sketches in JavaScript. Basic knowledge of JavaScript and Node.js will help you get the most out of this book.

What this book covers

Chapter 1, *Setting Up Your Development Environment*, this chapter covers the Raspberry Pi and how to use and set it up. This includes setting up raspbian on an SD card, installing Node.js, installing the Johnny-Five library, and installing the Raspi-IO library. The chapter will also explain the overarching concepts of Johnny-Five and Raspi-IO, as well as the benefits programming robotics in JavaScript brings.

Chapter 2, *Creating Your First Johnny-Five Project*, in this chapter, the reader will build their development environment for their Johnny-Five projects, and create their first Johnny-Five project: a blinking LED

Chapter 3, *Building Interactive Projects with RGB LED*, in this chapter, readers will be introduced to digital and PWM IO pins via LEDs–they will create a couple of projects with multiple LEDs and an RGB LED and explore fully the Johnny-Five LED API. We will also look at including outside Node.js libraries by including the color library to control the color of the RGB LED.

Chapter 4, *Bringing in Input with Buttons*, in this chapter, we'll show users how to incorporate basic input into their projects with buttons. Readers will learn to track the button by using Johnny-Five Button event.

Chapter 5, *Using a Light Sensor to Create a Night-Light*, in this chapter, we will add a sensor and create our first practical project: a night-light! The night-light will work by reading from the light sensor and if it's bright in the room, leave an LED off, but if it's dark, light it up! We'll also discuss the role sensors play in the robotics project ecosystem.

Chapter 6, *Using Motors to Move Your Project*, in this chapter, we will talk about making your project move with motors. This includes the extra hardware you will need for your project to be powered properly, how to wire motors to your project, the johnny-five motor API, and troubleshooting common problems when it comes to using motors with the Raspberry pi.

Chapter 7, *Using Servos for Measured Movement*, in this chapter, we will discuss measured movement in robotics projects with the servo, and build a servo that responds to a sensor. Readers will learn about servos and the Johnny-Five servo API, as well as build o project with one. They will also learn about the differences between servos and motors in the context of making robotics projects that move.

Chapter 8, *The Animation Library*, the animation library is a great way to fine-tune control over your johnny-five servo projects by controlling the speed, acceleration curve, and start and end points of your servo's movement. In this chapter, we'll look at the animation library and walk through how to control precision servo movements.

Chapter 9, *Getting the Information You Need*, in this chapter, we will look into why you would want to connect your NodeBots projects to the internet. We'll start by looking at ways you can use GET requests to obtain information from websites; like weather forecasts for your area. We'll build our first internet-connected bot using weather data and an RGB LED.

Chapter 10, *Using MQTT to Talk to Things on the Internet*, in this chapter, we'll talk about MQTT, a common IoT communications protocol. We'll look into using an MQTT broker, subscribing to it with our NodeBot, and using that data to react in real-time.

Chapter 11, *Building a NodeBots Swarm*, in the final chapter, we will go over possible paths to continue on your NodeBots journey. We will look at how to take the skills developed in the previous chapters and apply them to building multiple connected NodeBots, as well as exploring the avenues left to explore in the Johnny-Five library.

To get the most out of this book

You will require the following basic things in order to build all the projects that are included in the book:

- A laptop computer with any OS
- Raspberry Pi3
- Micro SD card (atleast 8 GB)
- Breadboard and wires
- Adafruit DC and Stepper Motor HAT for Raspberry Pi - Mini Kit
- Raspberry Pi 3 T-Cobbler
- GPIO expander board
- LCD hooked up to your Pi
- DC Toy / Hobby Motor - 130 Size
- 4 x AA Battery Holder with On/Off Switch

For more details on requirements, the setup of hardware and software is explained in each chapter of the book under the *Technical requirements* section.

Download the example code files

You can download the example code files for this book from your account at www.packtpub.com. If you purchased this book elsewhere, you can visit www.packtpub.com/support and register to have the files emailed directly to you.

You can download the code files by following these steps:

1. Log in or register at www.packtpub.com.
2. Select the **SUPPORT** tab.
3. Click on **Code Downloads & Errata**.
4. Enter the name of the book in the **Search** box and follow the onscreen instructions.

Once the file is downloaded, please make sure that you unzip or extract the folder using the latest version of:

- WinRAR/7-Zip for Windows
- Zipeg/iZip/UnRarX for Mac
- 7-Zip/PeaZip for Linux

The code bundle for the book is also hosted on GitHub at `https://github.com/PacktPublishing/Hands-On-Robotics-with-JavaScript`. In case there's an update to the code, it will be updated on the existing GitHub repository.

We also have other code bundles from our rich catalog of books and videos available at `https://github.com/PacktPublishing/`. Check them out!

Download the color images

We also provide a PDF file that has color images of the screenshots/diagrams used in this book. You can download it here: `https://www.packtpub.com/sites/default/files/downloads/HandsOnRoboticswithJavaScript_ColorImages.pdf`.

Conventions used

There are a number of text conventions used throughout this book.

`CodeInText`: Indicates code words in text, database table names, folder names, filenames, file extensions, pathnames, dummy URLs, user input, and Twitter handles. Here is an example: "Mount the downloaded `WebStorm-10*.dmg` disk image file as another disk in your system."

A block of code is set as follows:

```
board.on("ready", function() {
  // Everything else goes in here!
});
```

Any command-line input or output is written as follows:

```
sudo node implicit-animations.js
```

Bold: Indicates a new term, an important word, or words that you see onscreen. For example, words in menus or dialog boxes appear in the text like this. Here is an example: "To set up a feed, select **Feeds** in the left menu."

Warnings or important notes appear like this.

 Tips and tricks appear like this.

Get in touch

Feedback from our readers is always welcome.

General feedback: Email `feedback@packtpub.com` and mention the book title in the subject of your message. If you have questions about any aspect of this book, please email us at `questions@packtpub.com`.

Errata: Although we have taken every care to ensure the accuracy of our content, mistakes do happen. If you have found a mistake in this book, we would be grateful if you would report this to us. Please visit `www.packtpub.com/submit-errata`, selecting your book, clicking on the Errata Submission Form link, and entering the details.

Piracy: If you come across any illegal copies of our works in any form on the Internet, we would be grateful if you would provide us with the location address or website name. Please contact us at `copyright@packtpub.com` with a link to the material.

If you are interested in becoming an author: If there is a topic that you have expertise in and you are interested in either writing or contributing to a book, please visit `authors.packtpub.com`.

Reviews

Please leave a review. Once you have read and used this book, why not leave a review on the site that you purchased it from? Potential readers can then see and use your unbiased opinion to make purchase decisions, we at Packt can understand what you think about our products, and our authors can see your feedback on their book. Thank you!

For more information about Packt, please visit `packtpub.com`.

Setting Up Your Development Environment

1

Welcome! This book is designed to get you started with writing robotics code in JavaScript using the Raspberry Pi, Node.js, and the Johnny-Five framework. This chapter will fill in the details of what the Raspberry Pi is and how we're going to use it, and will also help you get your development environment ready.

The following topics will be covered in this chapter:

- What is the Raspberry Pi
- How we will use the Raspberry Pi
- Installing the operating system
- Setting up SSH and hardware interfaces
- Installing Node.js
- Installing Johnny-Five and Raspi-IO

Technical requirements

In order to get started, you'll need the following:

- **A Raspberry Pi 3**: Either the original or model B is fine.
- **A power supply**: Plugging the Raspberry Pi into a USB port on your computer can cause serious issues because it cannot supply enough power to allow the Raspberry Pi to function properly, so you'll need a proper wall wart power supply.
- **MicroSD card**: This needs to have at least 8 GB to hold the Raspbian OS and the code we're going to write. You'll also need a way to write to the SD card from your computer—either a full SD card adapter or a USB card reader.

- **A PC9685 GPIO expansion board**: There are expansion boards that require assembly on Adafruit (`https://www.adafruit.com/product/815`), but if you're not confident in your soldering, then there are plenty of preassembled ones available on Amazon if you search for `PC9685`.
- **Text editor**: Your code editor will be fine; we just need to edit a few files on the SD card once we've burned the OS image onto it.

If this is your first foray into a hardware project, I suggest getting a kit that contains at least the following items, as it will help you finish many of the projects in this book, and will provide you with the parts to create your own designs:

- Pi Cobbler
- Resistors
- LEDs
- A servo
- A motor
- Buttons
- Other sensors and peripherals

The following are some good examples of these items (at the time of writing):

- **The Raspberry Pi 3 B+ starter kit**: `https://www.sparkfun.com/products/14644`
- **The Adafruit Raspberry Pi 3 Model B Starter Pack**: `https://www.adafruit.com/product/2380`
- **If you already have a Pi, they sell the kit without the Pi as well**: `https://www.adafruit.com/product/3241`

What is the Raspberry Pi?

So, now you've got this green, credit-card-sized object with a bunch of ports that you recognize, and a bunch of pins, as shown in the following diagram. You can see some chips, and some parts you might not recognize. Before we talk about the power contained in this rather inconspicuous board, we need to clear up some vocabulary that we'll be using throughout the book:

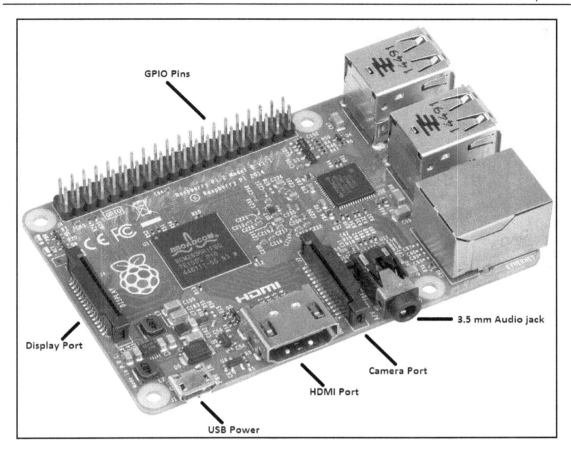

Microcontrollers

Microcontroller is a term that encapsulates a bunch of devices. It's a term used to describe a device that contains a processor, memory, and input/output peripherals (or ways to interact with those peripherals) that is meant for a particular type of task. One extremely common microcontroller is the Arduino Uno, and the Raspberry Pi technically falls into this category as well.

General-Purpose Input/Output (GPIO) pins

Microcontrollers interface with devices such as sensors, LEDs, and buttons using electrical signals that are sent and received through pins designed for input and/or output signals. These pins can be broken into multiple subcategories, as we'll find out in subsequent chapters, but you can address them as GPIO pins as a whole. We'll use that abbreviation throughout the book.

Debian and Raspbian

Debian is a distribution of Linux that is considered extremely user friendly for those new to using Linux. It contains many utilities that are commonly used while working with Linux preinstalled, and is compatible with a lot of the peripherals that you would use with a computer, such as Wi-Fi cards and USB devices.

Raspbian is a modified version of Debian specifically designed to run on Raspberry Pi devices. There are drivers for the GPIO pins, USB Wi-Fi devices, and expansion slots on the Pi that allow you to attach a specific display and camera.

There are two flavors of Raspbian—Raspbian Full and Raspbian Lite. Full has a graphical desktop with programs aimed at educational programming and development. Lite (which we will be using for the projects in this book) only has a command-line interface, but still has full functionality when it comes to Raspberry Pi peripherals. As of the time of writing, the current version of Raspbian is 4.14, nicknamed **Stretch**.

Johnny-Five and Raspi-IO

Back in 2012, Rick Waldron wrote a *node-serialport* program to operate an Arduino Uno with Node.js, and formed a library around it called Johnny-Five. Since then, the Johnny-Five library has grown to over 100 contributors, and can control over 40 platforms, including the Raspberry Pi! It can also control many kinds of sensors and peripherals that you can use to create the robotics project you've been dreaming of in Node.js!

One of the ways the Johnny-Five library has grown to support so many platforms is by creating what are called IO plugins. You create an IO plugin for each type of board you wish to control. For example, we will be installing and using the Raspi-IO plugin to use Johnny-Five with the Raspberry Pi.

What is great about this system is that the code you write in this book can be used on any other platform that Johnny-Five supports (you just need to change PIN numbers)! Writing code for Node.js botnets is much easier when you're using the same APIs for any devices you might use.

So, the Pi is technically a microcontroller...

Let's get back to the question of what the Raspberry Pi is. In short, it is a microcontroller. It has dozens of GPIO pins and can be used to interface with many physical peripherals in order to achieve specialized tasks. The low cost and small size allows the Raspberry Pi to be a versatile device, but the power involved allows you to use it for tasks that other microcontrollers may not pack the punch for.

...but it is also a computer!

An interesting fact about the Raspberry Pi is that, while it is a microcontroller, it can also be used as a fully fledged computer! While it certainly isn't the most powerful hardware, with full Raspbian installed, a Raspberry Pi attached to a monitor, keyboard, and mouse creates a great machine for kids and adults to learn programming on! The original intent of the Raspberry Pi was to create a low-cost educational machine to teach programming, and it exceeded every expectation in that regard. The fact that it's also a great microcontroller for the world of makers is a great bonus!

How we will use the Raspberry Pi

So, we've established that the Raspberry Pi is a very versatile and powerful machine for its size, but with so many options, it can be hard to figure out where to get started. Luckily, we have a plan that will walk you through your first Raspberry Pi and Johnny-Five projects so that you can keep up, but which will also empower you to build your way into advanced robotics projects.

Taking advantage of all that the Raspberry Pi has to offer!

The projects we will build will take advantage of the fact that the Raspberry Pi is both a microcontroller and a computer. We'll use the Linux operating system, via the Raspbian distribution, and leverage it to run our projects in Node.js. We'll also use Johnny-Five and Raspi-IO to leverage the GPIO of the Raspberry Pi in order to create robotics projects in a way that makes the code easy to understand and portable to many different hardware platforms.

Johnny-Five – letting us code hardware in Node.js

In the past, when you thought of robotics projects, it meant writing in C or C++, usually through the Arduino IDE and APIs. However, as microcontrollers have gotten more powerful, they are capable of running other programming languages, even scripting languages, such as subsets of Python and JavaScript.

And, of course, with computer/microcontroller hybrids, such as the Raspberry Pi, you're able to run stock Node.js, allowing you to create even advanced robotics projects without having to deal with any low-level languages. There are quite a few benefits to being able to code robotics projects in Node.js:

- **Event-based systems**: In the Arduino and C/C++ level of robotics programming, you will need to check the state of everything through each iteration of a loop, and act accordingly. This can create monolithic functions and code paths. With Node.js and Johnny-Five, we can use event emitters and systems, which fit in surprisingly well as they can read sensors and interact with peripherals in the real world, where things take time. This will help you to organize code in a way that reflects the asynchronous way the world works.
- **Garbage collection/automatic memory management**: While Arduino and C++ handle most memory management for you, programming in microcontrollers that use C requires strict memory management. While you may need to bear the resource constraints of the Raspberry Pi in mind from time to time, it is much easier than the days of 20K SRAM.

- **Using a language you already know**: Instead of trying to remember the way things work in a new language, it will accelerate your learning in the field of electronics and robotics if you focus on learning fewer things at once. Using Node.js can help you focus on learning the wide and varied world of electronics, instead of adding on the extra work of remembering whether it's `uint8_t` or `uint16_t`.

Installing the operating system

In order to get started with Johnny-Five and the Raspberry Pi, we will need to set up the Raspbian operating system by burning it to a microSD card. Then, we'll need to edit some files on the SD card in order for the Raspberry Pi to boot with Wi-Fi and the ability to SSH in. Lastly, we'll need to boot up the Raspberry Pi and get some settings in place before finally installing Node.js, Johnny-Five, and Raspi-IO.

Downloading Raspbian Lite

The following steps will show you how to download Raspbian Lite:

1. The first step is downloading the Raspbian Lite image so we can burn it to our microSD card. The best place to get the image from is `https://www.raspberrypi.org/downloads/raspbian/`, as shown in the following screenshot:

A screenshot of the Raspbian download page, with both Full and Lite download links

2. Select **RASPBIAN STRETCH LITE** (or whichever version is the current one), which will replace the word **STRETCH**. Give yourself some time for this step to complete; although Raspbian Lite is much smaller than Raspbian Full, it is still several hundred megabytes, and can take time to download!

If you're preparing to run a class, hackathon, or some other event using the Raspberry Pi and Raspbian, it's best to predownload it and place it on a flash drive to hand around, as conference and event Wi-Fi can be a bit slower than normal, or even drop in and out, so be prepared!

Burning the image to an SD card

Luckily, the tools for burning OS images to SD cards have evolved arcane command-line tools that can overwrite your computer's hard drive as easily as it can the SD card. My current favorite is called **Etcher**, and it can be downloaded for any platform at `https://etcher.io/`, as shown in the following steps:

1. The free version is more than enough for our needs, so download and install it as you download Raspbian Lite.
2. Once they are both downloaded, you'll want to place the micro SD card in your computer, either by placing it into a full-sized SD adapter and then into a slot on your computer, or by using a USB-to-micro-SD adapter. Whichever you use, make sure your computer can see the volume as a drive before continuing. Then, boot up Etcher. The following screenshot shows Etcher running on a Mac:

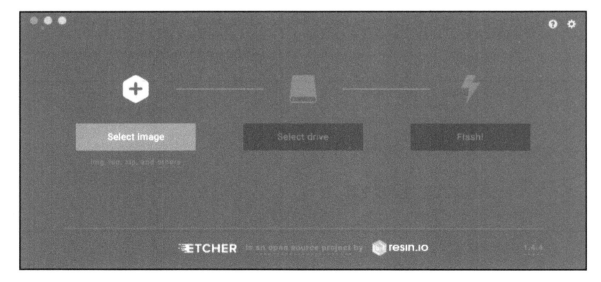

A screenshot of the Etcher program running on a Mac

3. Once you see a window similar to the preceding screenshot, you'll need to select the Raspbian Lite image you just downloaded. You don't even need to unzip the .zip file—Etcher can handle that outright! Once you've selected the image, Etcher should select your micro SD card drive, so long as your machine can see it! Once you've ensured that the image and micro SD card drive are properly selected, hit **Flash!** to begin the process.

Sometimes, with larger micro SD cards, you'll get a warning from Etcher about the drive being very large (this happens to me when I use 64 GB cards). This is to prevent you from overwriting your computer's hard drive. You can bypass the warning by going through a confirmation window—just be absolutely sure that your micro SD card drive is selected first!

A few minutes will pass as Etcher burns the image to your micro SD card, then verifies that it is present on the card. Once this is done, **remove and reinsert the micro SD card so that your computer recognizes it as a drive again**; the drive should be named boot. We're not quite done editing the image files yet, so Etcher's polite attempt to eject the micro SD card drive needs to be ignored.

Editing files on the SD card

We need to edit and create some files on our Raspberry Pi's image in order to be able to access it with SSH when we turn on the Raspberry Pi. First, we'll set up the Wi-Fi using the following steps:

If you're using an Ethernet cable and port to connect the Raspberry Pi to the internet, you can skip this step. If this doesn't get the Wi-Fi to work, you'll want to look at the information box under the section *Booting up the Pi* for troubleshooting steps and an alternative (if clunkier) way to set this up.

1. In order to set up the Wi-Fi, you'll want to create a file in the root of the micro SD card drive called wpa_supplicant.conf that contains the following text:

```
ctrl_interface=DIR=/var/run/wpa_supplicant GROUP=netdev
update_config=1
network={
    ssid="yourNetworkSSID"
    psk="yourNetworkPasswd"
}
```

2. Replace `yourNetworkSSID` with the SSID of the Wi-Fi network you wish to connect to, and then replace `yourNetworkPasswd` with that Wi-Fi's network password.

 The Raspberry Pi's Wi-Fi chip can only connect to 2.4 GHz networks at the time of writing, so you need to make sure that you input a network that operates on that bandwidth or your Raspberry Pi will not be able to connect!

3. After you've set up the Wi-Fi network, you'll want to tell the Raspberry Pi to allow you to SSH into it. To do this, you'll want to create a file called `ssh` in the same `root` folder as the `wpa_supplicant.conf` file. Make sure that the `ssh` file is empty and has no extension. When you're all done, the `root` directory of the micro SD card drive will look similar to the following screenshot:

A list of files on the micro SD drive, once I've made my edits

Once this is all done, fully eject the microSD drive, take the microSD card and insert it into the Raspberry Pi. We're now ready to boot up the Raspberry Pi and start installing software.

Booting up the Pi

Once the micro SD card is inserted, plug your power source into your Raspberry Pi. A red and green LED should light up. The red LED should be solid—this is the power indicator. The green light will flicker on and off—this is the LED that indicates activity. The Raspberry Pi should take a maximum of a minute or so to boot up once you've plugged it in. Next, we'll SSH in.

SSHing from a Linux or Mac

If you're on a Mac or Linux machine, you'll open up a Terminal and type the following:

```
ssh pi@raspberrypi.local
```

If you're successful, you'll see a question appear, asking about the authenticity of the host. Respond by typing `yes` and hitting *Enter*. You'll then be asked for a password, which is `raspberry`, as shown in the following screenshot:

```
Code/bots/book
➜ ssh pi@raspberrypi.local
The authenticity of host 'raspberrypi.local (2600:1700:211:3920:6b7a:ca9e:9c9e:1cee)' can't be established.
ECDSA key fingerprint is SHA256:/SVFAnCYjqbNacZwuALCgTUYf+oZXsGEpSj5oHRNiZQ.
Are you sure you want to continue connecting (yes/no)? yes
Warning: Permanently added 'raspberrypi.local,2600:1700:211:3920:6b7a:ca9e:9c9e:1cee' (ECDSA) to the list of known hosts.
pi@raspberrypi.local's password:
```

Successful SSH into a Raspberry Pi from a Mac Terminal

Once you've entered the password, you should see the following:

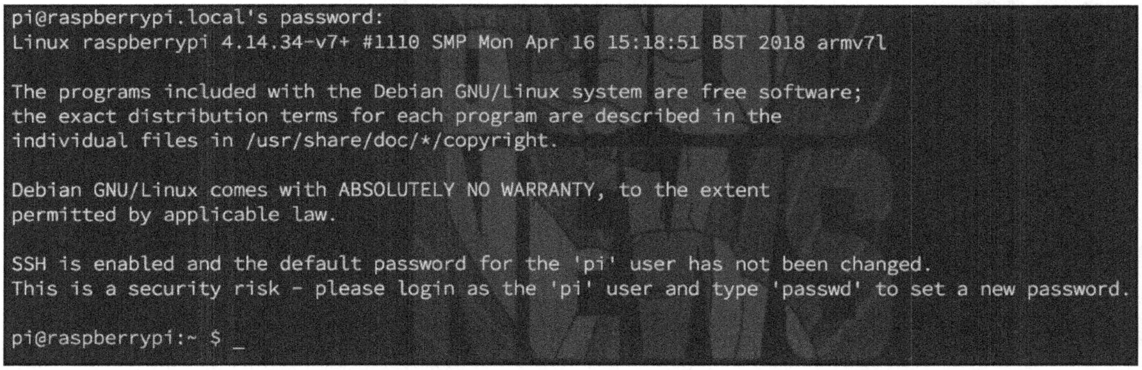

```
pi@raspberrypi.local's password:
Linux raspberrypi 4.14.34-v7+ #1110 SMP Mon Apr 16 15:18:51 BST 2018 armv7l

The programs included with the Debian GNU/Linux system are free software;
the exact distribution terms for each program are described in the
individual files in /usr/share/doc/*/copyright.

Debian GNU/Linux comes with ABSOLUTELY NO WARRANTY, to the extent
permitted by applicable law.

SSH is enabled and the default password for the 'pi' user has not been changed.
This is a security risk - please login as the 'pi' user and type 'passwd' to set a new password.

pi@raspberrypi:~ $ _
```

Successful login to a Raspberry Pi via SSH from a Mac

SSHing from Windows

In order to SSH from a Windows machine, you'll need to use a program called PuTTY. You can get it on `https://putty.org/`. But first, you'll want your Raspberry Pi's IP address. You'll need a monitor, an HDMI cable, and a USB keyboard. Once you have these, go through the following steps:

1. Plug the monitor, HDMI cable, and USB keyboard into the Raspberry Pi **before booting it up**. Then plug in the power supply.
2. When it prompts you for a username, type `pi`. When it asks for a password, enter `raspberry`. Once you are logged in, type `ifconfig`. You should see a lot of information appear.
3. Look for the `wlan0` section and the `inet` address under that heading. For the following output, the IP is `192.168.1.106`, as shown in the following screenshot. Write this IP down. Then, you can unplug the display and the keyboard—you won't need them again:

```
pi@raspberrypi:~ $ ifconfig
eth0: flags=4099<UP,BROADCAST,MULTICAST>  mtu 1500
        ether b8:27:eb:5a:56:f8  txqueuelen 1000  (Ethernet)
        RX packets 0  bytes 0 (0.0 B)
        RX errors 0  dropped 0  overruns 0  frame 0
        TX packets 0  bytes 0 (0.0 B)
        TX errors 0  dropped 0 overruns 0  carrier 0  collisions 0

lo: flags=73<UP,LOOPBACK,RUNNING>  mtu 65536
        inet 127.0.0.1  netmask 255.0.0.0
        inet6 ::1  prefixlen 128  scopeid 0x10<host>
        loop  txqueuelen 1000  (Local Loopback)
        RX packets 0  bytes 0 (0.0 B)
        RX errors 0  dropped 0  overruns 0  frame 0
        TX packets 0  bytes 0 (0.0 B)
        TX errors 0  dropped 0 overruns 0  carrier 0  collisions 0

wlan0: flags=4163<UP,BROADCAST,RUNNING,MULTICAST>  mtu 1500
        inet 192.168.1.106  netmask 255.255.255.0  broadcast 192.168.1.255
```

Getting the IP address from the Terminal

4. Once you have the IP address for your Raspberry Pi, you can boot up PuTTY. The window that opens is the configuration window, as shown in the following screenshot:

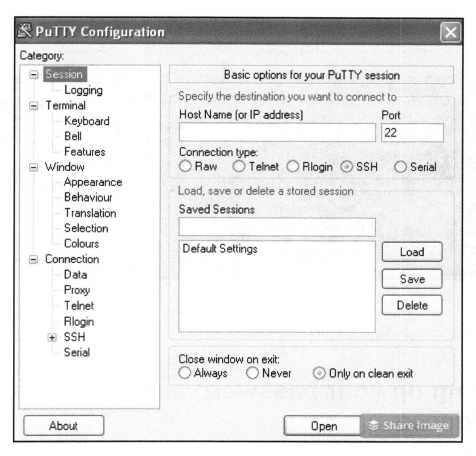

The PuTTY configuration window

5. Type the IP address that you obtained into the field labeled **Host Name (or IP address)** and click the **Open** button. You'll be asked about the authenticity of the host (only the first time you connect). Select **Yes**. Then enter `Pi` as the username and `raspberry` as the password when prompted. Once that's done, you should see the following:

Successful login to the Raspberry Pi with PuTTY

Now that everyone's logged in, let's set up our Raspberry Pi for our projects!

Setting up your password and hardware interfaces

Now that we have our Raspberry Pi connected to the Wi-Fi and we're SSHed in, we need to make a few changes before we install Node.js and get started with our coding.

First things first – change your password!

When you log in, your Raspberry Pi will warn you that having SSH enabled with the default username and password isn't very secure, and it's absolutely right! The first step is to change our password.

In order to do so, in your SSH window, type in `passwd` and hit *Enter*. You'll be prompted for your current password (`raspberry`) and a new password. Type in whatever you like (just don't forget it)! You'll be asked to confirm it, and voila! The new password is set, as shown in the following screenshot. Your Raspberry Pi will be much more secure:

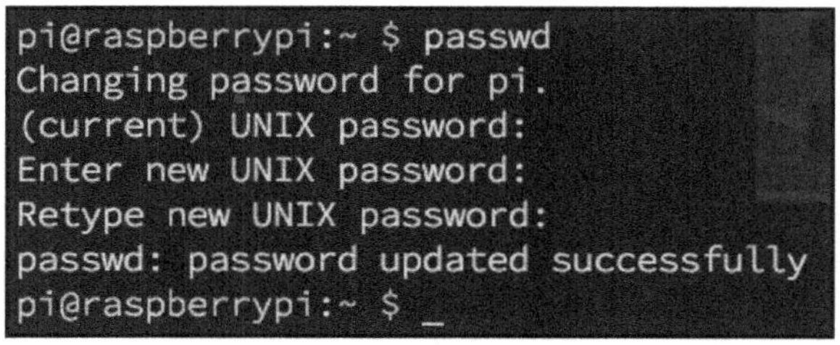

```
pi@raspberrypi:~ $ passwd
Changing password for pi.
(current) UNIX password:
Enter new UNIX password:
Retype new UNIX password:
passwd: password updated successfully
pi@raspberrypi:~ $
```

Changing your Pi password

Updating the Raspberry Pi

Next, you'll make sure that the Raspberry Pi is updated and ready to go by running the following command:

```
sudo apt-get update && sudo apt-get upgrade
```

This will take a while, but it's worth it to make sure everything is properly updated.

Turning on the hardware interfaces

Next, we'll set up the Raspberry Pi so that our hardware code can run. Run the following command:

```
sudo raspi-config
```

You'll be greeted with a graphical menu with lots of different options, as shown in the following screenshot:

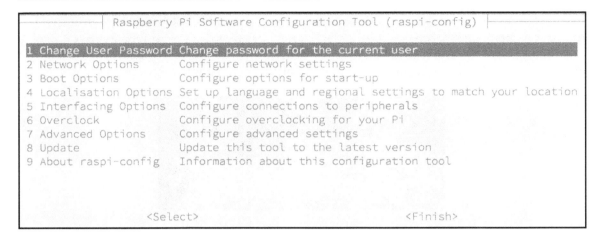

The raspi-config menu

You'll want to use the arrow keys to select Interfacing Options, and then select I2C and Yes to turn it on. Repeat for SPI, then use *Tab* to close the menu. When it prompts you to reboot, say Yes, then SSH back in, because you're ready to install Node.js, Johnny-Five, and Raspi-IO!

Installing Node.js, Johnny-Five, and Raspi-IO

So, now that our Raspbian OS is installed and set up, it's time to install Node.js (which comes bundled with npm), Johnny-Five, and Raspi-IO!

Installing Node.js and npm

In days past, you would have to compile the Node.js source on your Pi, to varying degrees of success because of the nonexistence of binaries for the ARM processor that Raspberry Pi uses. Luckily now, because of a rousing amount of third-party support in the past few years, you can easily download the binary from the https://nodejs.org/en/ website! But how are we going to do this from the command line of our Raspberry Pi?

Detecting your version of ARM processor

If you're using the Raspberry Pi 3 Model B recommended by this book, you're most likely on ARM v8 (the Raspberry Pi 3 original is ARMv7, which is fine too!). But you should always double-check (doubly so if you're using a different Raspberry Pi, such as the Pi Zero or Pi 2/1 series). To check the ARM version on your Raspberry Pi, run the following in your SSH Terminal:

```
uname -m
```

You'll see a return message that looks like `armv#`, where # is a number (possibly followed by a letter). That number is what is important, because that number tells us which Node.js binary we will need. Once you have your ARM version, go through the following steps:

1. Head to the Node.js download page at `https://nodejs.org/en/download/`, as shown in the following screenshot:

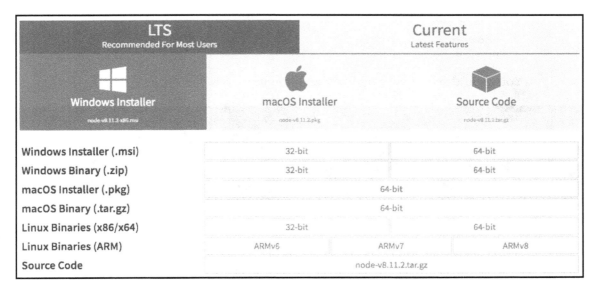

	LTS Recommended For Most Users		Current Latest Features
	Windows Installer node-v8.11.2-x86.msi	macOS Installer node-v8.11.2.pkg	Source Code node-v8.11.2.tar.gz
Windows Installer (.msi)	32-bit		64-bit
Windows Binary (.zip)	32-bit		64-bit
macOS Installer (.pkg)	64-bit		
macOS Binary (.tar.gz)	64-bit		
Linux Binaries (x86/x64)	32-bit		64-bit
Linux Binaries (ARM)	ARMv6	ARMv7	ARMv8
Source Code	node-v8.11.2.tar.gz		

A snapshot of the Node.js binary download page

2. Right-click on the ARM version link you need and copy the URL. Then, run the following in your Raspberry Pi's SSH Terminal:

```
wget <binary-download-url>
```

Replace `<binary-download-url>` (carats, too!) with the URL you copied from the Node.js download website. Once it's downloaded, we need to extract the archive using the following code:

```
tar -xvf node-v****-linux-armv**.tar.xz
```

3. The asterisks will differ depending on the current LTS version of Node.js and your ARM version. The Raspberry Pi will spit out a lot of filenames to the console, then give you back your shell prompt. This means that the binaries have been extracted into your `home` folder. We need to place them into the `/usr/local` folder. To do that, run the following:

```
cd node-v****-linux-armv**
sudo mv ./lib/* /usr/local/lib
sudo mv ./share/* /usr/local/share
```

4. This will move all of the precompiled binaries to their new homes on your Raspberry Pi. Once this is done, run the following:

```
node -v
npm -v
```

You should see something like the following:

```
pi@raspberrypi:~/node-v8.11.2-linux-armv7l $ sudo mv ./lib/* /usr/local/lib
pi@raspberrypi:~/node-v8.11.2-linux-armv7l $ sudo mv ./share/* /usr/local/share
pi@raspberrypi:~/node-v8.11.2-linux-armv7l $ node -v
v8.11.2
pi@raspberrypi:~/node-v8.11.2-linux-armv7l $ npm -v
5.6.0
```

Successful Node.js installation results

5. If that's all well and good, you now have Node.js and npm installed! Let's wrap this up with Johnny-Five and Raspi-IO! Note that you can absolutely clean up the binary downloads by running the following:

```
cd ~
rm -rf node-v**-linux-armv**
rm -rf node-v****-linux-armv**.tar.xz
```

Some of you with more Debian experience may be asking, *well, why can't we just use apt-get*? The short answer is that the package with the name `node` was taken a very long time ago, and because that is the case, and because `sudo apt-get install nodejs` is outdated (at the time of writing, using this command will install `v4` when we need `v8+`, if it installs Node.js at all), we need to download the binaries and move them ourselves.

Installing Johnny-Five and Raspi-IO

To install Johnny-Five, once you've made sure Node.js and npm are installed, run the following command:

```
npm i -g johnny-five raspi-io
```

This installs the libraries globally; you won't have to reinstall it every new project. And that's it! You're ready to start developing Node.js robotics projects on the Raspberry Pi with Johnny-Five!

Summary

It feels like a lot, but you've now completed everything you need to have a fully-fledged development environment for the projects in this book, and you've taken your first steps toward building robots with JavaScript. You've learned more about what the Raspberry Pi is and why we're using it, and how to get the operating system image ready to go!

Questions

1. What is the common operating system for the Raspberry Pi that we'll be using in the projects in this book, and what Linux distribution is it based on?
2. What does GPIO stand for?
3. Who originally started the Johnny-Five project, and what did they use it to control?

4. What command do you run on the Raspberry Pi to find out what ARM architecture it uses?
5. Why is changing the default Raspberry Pi password important?
6. What are two benefits of using JavaScript and Node.js for robotics code?
7. Why do we have to download the Node.js binaries instead of using Raspbian's package manager?

Further reading

You can use the following sources for further reading relating to the topics covered in this chapter:

- **Learn more about the Raspberry Pi from the website of the Raspberry Pi organization**: https://www.raspberrypi.org/
- **Learn more about Johnny-Five from the main project page for Johnny-Five (we'll be seeing a lot of this site as we use their documentation to complete many of the book's projects)**: http://johnny-five.io/
- **Learn more about the Raspbian operating system from the Raspbian website**: https://www.raspberrypi.org/documentation/raspbian

2
Creating Your First Johnny-Five Project

Now that we've set up our development environment, it's time to start writing code and making LEDs light up! We'll start by running the *Hello World!* of Johnny-Five robotics: making an LED blink. In the process, we'll look at how to navigate the Johnny-Five and Raspi-IO API documents, and examine the event system in Johnny-Five.

The following topics will be covered in this chapter:

- Creating a project folder
- Installing Johnny-Five and Raspi-IO
- Wiring up an LED
- Making an LED blink

Technical requirements

You'll need the Raspberry Pi that you set up in Chapter 1, *Setting Up Your Development Environment*, a breadboard, and a Pi Cobbler for easier pin access. You can get a Pi Cobbler (also sometimes called a Pi Wedge) from Adafruit, SparkFun, or Amazon. A Pi Cobbler also comes in the kits recommended in Chapter 1, *Setting Up Your Development Environment*.

 The example code for this chapter is here: https://github.com/ PacktPublishing/Hands-On-Robotics-with-JavaScript/tree/master/ Chapter02.

The following diagram shows two different Raspberry Pi Cobblers, both from Adafruit. The one on the right has the ribbon cable attached:

We'll talk about how to set up the cobbler later in this chapter. You'll also need an LED, some jumper or breadboard wires, and a 330-ohm resistor.

In case you're asking yourself *what's a resistor, and what does it do?*, the short explanation is that a resistor will prevent the 5V electricity from the pin from burning out your LED, which needs closer to 3.3V of electricity. For a better primer on electricity, voltage, and resistors, there is some great, free material on SparkFun's website. You can access this material via the following links:

Electricity: https://learn.sparkfun.com/tutorials/what-is-electricity

Resistors: https://learn.sparkfun.com/tutorials/resistors

Creating a project folder

I find the best way to organize your Raspberry Pi is to put each project in its own folder. In the source code that accompanies this book, I've done just that. But let's walk through how to set up your own project folders. First, you'll want to create the folder itself. For the project in this chapter, which we'll call led-blink, you'll want to run the following:

```
cd ~
mkdir led-blink
```

Make sure that you're running this in your SSH session to the Raspberry Pi, and not on your desktop.

 From here on out, unless the text directly says to run something on your desktop, you should run all of your commands in the SSH session to your Raspberry Pi that we set up in `Chapter 1`, *Setting Up Your Development Environment*.

Setting up npm to manage our modules

We're going to be using more than just Johnny-Five and Raspi-IO to create our projects, and you want to be able to move your code around via your favorite Git hosting service, perhaps to move it to a new Raspberry Pi, for example. In order to make this as smooth as possible, we're going to make sure that `npm` knows how to accurately recreate your projects. For this, we want a preprepared `package.json` file. To do this, navigate into your `project` folder and tell it to initialize:

```
cd led-blink
npm init -y
```

 The `-y` in the `npm init` command tells `npm` to *use the default answer to all initialization questions*. This is fine for projects that only you will use, but if you plan to deploy your work for others to use, or create your own `npm` modules, be sure to edit your `package.json` accordingly.

These commands create our `package.json` so that when we install `npm` modules with `--save`, the manifest will update so that when you move your project, there's a complete record of our dependencies.

Getting started with Johnny-Five and Raspi-IO

Now that our project folder is ready for dependencies, we'll start exploring the Johnny-Five and Raspi-IO documentation that'll help us create the projects in this book.

Gathering resources and documentation

There are two main sources of documentation that we'll be using for the projects in this book:

- **The Johnny-Five website**: http://johnny-five.io/
- **The Raspi-IO GitHub README and wiki**: https://github.com/nebrius/raspi-io

The following screenshot shows the Raspi-IO README on GitHub:

README.md

Raspi-io

gitter join chat

Raspi-io is a Firmata API compatible library for Raspbian running on the Raspberry Pi that can be used as an I/O plugin with Johnny-Five. The API docs for this module can be found on the Johnny-Five Wiki, except for the constructor which is documented below. Raspi IO supports all models of the Raspberry Pi, except for the Model A.

If you have a bug report, feature request, or wish to contribute code, please be sure to check out the Contributing Guide.

System Requirements

- Raspberry Pi Model B Rev 1 or newer (sorry Model A users)
- Raspbian Jessie or newer

We will be using the Johnny-Five documentation at `johnny-five.io` to look up API calls and other information about the Johnny-Five library, and the Raspi-IO README for Raspberry Pi-specific information, including pin numbers.

Taking a look at the LED-blink project

The first thing we'll need is from the Raspi-IO README: we're going to read and run their `led-blink` code as our *Hello World!* Let's take a look at the code as a block:

```
const Raspi = require('raspi-io');
const five = require('johnny-five');

const board = new five.Board({
  io: new Raspi()
});

board.on('ready', () => {

  // Create an Led on pin 7 (GPIO4) on P1 and strobe it on/off
  // Optionally set the speed; defaults to 100ms
  (new five.Led('P1-7')).strobe();

});
```

This doesn't look like much, but there's a lot going on here! The first two lines use `require` to pull in the `johnny-five` and `raspi-io` modules. Then, we begin constructing a `board` object, and we pass a new instance of the `raspi-io` module in as its I/O. This is how we tell the Johnny-Five library we're running this code on a Raspberry Pi.

Next, we'll set up an event listener on our board object on the `ready` event. According to the Johnny-Five documentation, this event fires *when the board instance object has completed any hardware initialization that must take place before the program can operate.* This means you shouldn't run any robotics-related code outside of this event handler, because you cannot be sure your board is ready to receive hardware commands.

The comments starting on line 10 are very helpful, as they tell us where to hook up our LED. We'll be using pin 7 (GPIO 4)—that means the seventh physical pin from the top, which is designated as GPIO 4.

Raspberry Pi pin numbers

Pin 7... is labeled GPIO 4? That's confusing! Luckily, there are many pin diagrams freely available to help us translate, as shown in the following diagram:

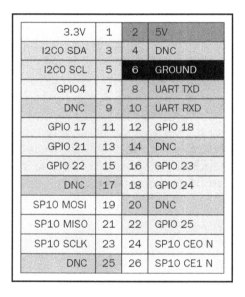

A GPIO/pin map (source: raspberrypi.org)

Also, the Raspi-IO library will accept many names for the same pin, as shown in the handy conversion table in the wiki:

Model A+/B+/Raspberry Pi 2/Raspberry Pi 3/Raspberry Pi Zero						
P1 Header						
Physical Pin	Wiring Pi Pin	Peripherals		Peripherals	Wiring Pi Pin	Physical Pin
P1-1		3.3V		5V		P1-2
P1-3		SDA0		5V		P1-4
P1-5		SCL0		GND		P1-6
P1-7	7	GPIO4		GPIO14/TXD0	15	P1-8
P1-9		GND		GPIO15/RXD0	16	P1-10
P1-11	0	GPIO17		GPIO18/PWM0	1	P1-12
P1-13	2	GPIO27		GND		P1-14
P1-15	3	GPIO22		GPIO23	4	P1-16
P1-17		3.3V		GPIO24	5	P1-18
P1-19	12	GPIO10/MOSI0		GND		P1-20
P1-21	13	GPIO9/MISO0		GPIO25	6	P1-22
P1-23	14	GPIO11/SCLK0		GPIO8/CE0	10	P1-24
P1-25		GND		GPIO7/CE1	11	P1-26
P1-27		Do Not Connect		Do Not Connect		P1-28
P1-29	21	GPIO5		GND		P1-30
P1-31	22	GPIO6		GPIO12/PWM0	26	P1-32
P1-33	23	GPIO13/PWM1		GND		P1-34
P1-35	24	GPIO19/MISO1/PWM1		GPIO16	27	P1-36
P1-37	25	GPIO26		GPIO20/MOSI1	28	P1-38
P1-39		GND		GPIO21/SCLK1	29	P1-40

The pin table from the Raspi-IO wiki

Keeping one of these pin guides handy is helpful when wiring up any Johnny-Five project on the Raspberry Pi.

Wiring up an LED

Now that we've gone through the documentation and figured out what goes where, we can start assembling our Raspberry Pi project. You'll need your Raspberry Pi, Pi Cobbler, two breadboard wires, an LED (doesn't matter what color), and a 300-ohm resistor.

Putting together and attaching the cobbler

In order to make sure the cobbler is seated correctly, you'll want to make sure that the ribbon cable points outward from the Raspberry Pi when placed on the GPIO pins, and that the little tab on the side of the connector faces the right way in the cobbler itself (this is usually ensured by a plastic wall around the pins that the ribbon cable plugs into; make sure that you check that it's lined up before applying too much pressure!).

You'll want to seat the cobbler on a breadboard that's at least half sized, though I tend to prefer full size for Raspberry Pi projects. Make sure the two rows of pins on the cobbler are on opposite sides of the groove down the center of the breadboard, as shown in the following photograph:

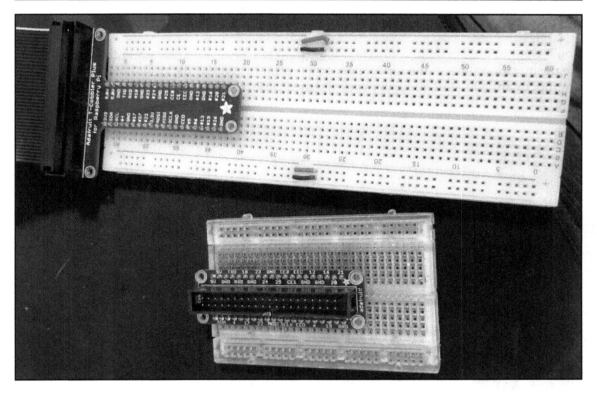

Cobblers on a full-size (top) and half-size (bottom) breadboard

 New to breadboards? There's a great explanation of how they work (as well as some neat trivia) on the SparkFun website (`https://learn.sparkfun.com/tutorials/how-to-use-a-breadboard`).

Attaching the resistor and LED

You'll want to use a wire to connect GPIO 4 (pin 7) to a 330-ohm resistor and the resistor to the positive (long) leg of the LED. Then, you'll want to connect the negative (short) leg to ground, or any pin marked GND. Your finished project will look something like this:

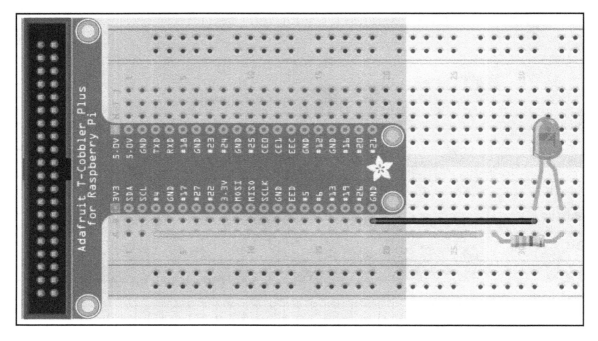

Your LED project

Now that your LED is wired up, it's time to make it blink!

Making the LED blink

In order to make the LED blink, we'll need to install the code on the Raspberry Pi, and then run it!

Putting your code on the Raspberry Pi

If you wrote your code on your desktop and need to transfer it to your Raspberry Pi, there are a couple of ways to go about it: you can use `rsync` on an macOS X or Linux machine:

```
rsync ./blink-led.js
pi@raspberrypi.local:~/<project folder>/
```

Replace `<project folder>` with the folder you want to transfer into (for example, the book folder would be `hands-on-robotics-with-javascript/ch2/blink-led`).

 For Windows, follow the guide for installing and using WinSCP at `https://winscp.net/eng/docs/ui_commander`.

Running your code

Once your code is on your Raspberry Pi, you'll want to switch to your SSH session and run the following, if you're using the source code from this book:

```
cd hands-on-robotics-with-javascript/ch2/blink-led
```

Otherwise, use `cd` to enter the folder where you stored your `blink-led.js` file on the Raspberry Pi. Then, run the following:

```
sudo node blink-led.js
```

Note that the Raspi-IO plugin requires you to run the command as `sudo`. If all has gone well, you should see a quickly blinking LED on your breadboard. If not, here are some troubleshooting steps:

1. Check the wiring
2. Double-check the wiring (seriously, 95% of the time it's a wiring issue)
3. Make sure that the Node.js script did not experience an error on the Raspberry Pi

Summary

Congratulations! You've made a Johnny-Five bot! In this chapter, you've learned how to wire up an LED, navigate the documentation for Johnny-Five and Raspi-IO, and run your code on the Raspberry Pi!

Questions

1. Look in the Johnny-Five documentation, under the LED heading in the section on the API. Look for the `strobe` function. What does the first argument do? What would happen if you passed 500 as that first argument?
2. What is the second argument in the `LED.strobe()` function? How would this come in handy for applications waiting for the LED to be off?
3. Does the Johnny-Five LED object emit any events? Why, or why not?
4. Using the Raspi-IO documentation, what does the Raspberry Pi pin P1-29 translate to in terms of GPIO #?
5. Using the Johnny-Five documentation, name a function that is an alias for the `LED.strobe()` function.
6. What happens before the board's `ready` event fires in a Johnny-Five application?

Further reading

You can consult the following sources for further reading related to the topics covered in this chapter:

- **The Johnny-Five documentation**: `johnny-five.io`
- **The Johnny-Five GitHub repository**: `https://github.com/rwaldron/johnny-five`
- **The Raspi-IO library**: `https://github.com/nebrius/raspi-io`

3
Building Interactive Projects with RGB LED

Now that we've built a project with Johnny-Five and Raspi-IO, it's time to tackle GPIO expanders and PWM outputs, and build an interactive project with an RGB LED. We'll also learn more about the Johnny-Five REPL, learn how a PWM pin works, and use this knowledge to control an RGB LED from the command line.

The following topics will be covered in this chapter:

- Looking at the LED and LED.RGB API
- PWM pins and GPIO expanders
- Bringing in other node packages with color
- The Johnny-Five REPL

Technical requirements

You will have already installed all the software prerequisites for this chapter from Chapter 1, *Setting Up Your Development Environment*. You'll want to make sure your Raspberry Pi is connected to the internet and that you have SSHed in using your method of choice.

 The example code for this chapter can be found at https://github.com/nodebotanist/hands-on-robotics-with-javascript/tree/master/ch3.

As for hardware, you will need the following:

- Your Raspberry Pi
- Cobbler/breadboard
- Breadboard wires
- PCA9685 GPIO expander board
- RGB LED
- 330-ohm resistor x 3

Looking at the LED and LED.RGB API

We took a brief look at the LED API in Johnny-Five in the last chapter, but in this chapter, we will delve deeper and talk about the PWM output and the cousin of the standard LED, the RGB LED—so named because it has a red, green, and blue channel, and can replicate thousands of colors. We will use an RGB LED, as well as some of the more powerful tools built into Johnny-Five, to build an interactive project in this chapter.

The LED object

The LED object is usually the first thing in Johnny-Five that people look through the documentation for. It's also a great object to use to outline the general structure of the object documentation. Let's take a look at each section and get a grasp of where we should look for what later:

- **Parameters**: This section addresses the parameters that need to be passed into the object constructor, and what form they need to be in (order, object key, and so on).
- **Shapes**: These are the fields attached to the constructed object that may be useful to the user in writing their code. They can be read-only, and are marked if this is the case.
- **Component Initialization**: This is usually a piece of sample code, but it's always a description of how to construct a common usage of the object in question. If there are multiple controllers for a specific component, they are enumerated with examples for each controller. This will come in handy for our GPIO expander board.

- **Usage**: This is a sample code denoting how to use the most basic functions of an object; for the LED, this is the `blink` function, and for sensors, this would show the function that you would usually use to get readings from the sensor.
- **API**: This is a full documentation of every function available to the object, including the parameters and intended result.
- **Events**: Many objects emit events (such as the board's `ready` event); this section details when they will fire.
- **Examples**: The Johnny-Five community is a fantastic source of examples, and the examples that are relevant to the object in question will be cataloged and linked in this section.

Take a moment to get used to the LED documentation (`http://johnny-five.io/api/led/`), because the `Led.RGB` object is essentially a subclass of the LED object, and will inherit many of its functions.

The Led.RGB object

Once you've acquainted yourself with the LED object, click on the **Led.RGB** link in the sidebar, as shown in the following screenshot:

You'll be taken to the Led.RGB documentation page. In the *Component Initialization* section, look for the *LED RGB PCA9685* section. Ignore the wiring diagram (it's for the Tessel 2, a different microcontroller), but do make a note of the example code, as shown here:

```
new five.Led.RGB({
    controller: "PCA9685",
    pins: {
      red: 2,
      green: 1,
      blue: 0
```

```
    }
});
```

This is the code that we will use to initialize our RGB.LED object.

We will also need the API section in order to determine the functions and parameters that we will need for using the RGB.LED object. Take a look at the color() function, in particular.

Now that we have our module for converting colors into RGB values, we can start talking about how to use an RGB LED in order to get those colors into our projects.

PWM pins and GPIO expanders

Before we wire up and run our RGB LED project, a discussion about PWM pins and GPIO expanders is warranted, because these are topics that will affect most Johnny-Five projects that you will complete.

How do PWM pins work?

You don't always want an LED to at its full brightness, especially in the case of RGB LEDs, where the brightness of each channel (red, green, and blue) determines the perceived color of the LED. The pins on most microcontrollers are digital: they are either HIGH at 5V or LOW at 0V. So how do you adjust the brightness of an LED with these types of pins? The answer involves the idea of average voltage and the speed at which we can flip a digital pin from HIGH to LOW.

Pulse-width modulation, or PWM, pins operate by setting, effectively, the percentage of time that a pin is HIGH and LOW. The following screenshot shows an oscilloscope reading for the state of a pin running at 50% over a short period of time:

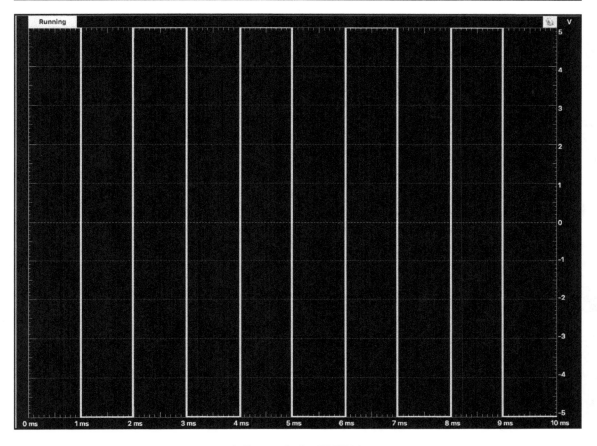

Oscilloscope reading for a 50% PWM pin

Instead of the LED flickering on and off, this results in the LED appearing to glow at half-brightness: this is because the human eye cannot keep up with the speed of the state changes, and sees the LED as on, but dimmed. This helps us to use RGB LEDs to create thousands of different colors by combining a red, green, and blue channel at varying degrees of brightness.

When we set the colors in our Johnny-Five code, we can pass values from 0 to 255 for red, green, and blue. This works well with web hex colors, which use the same range. In general, you can set a PWM pin from 0 to 255 (with some exceptions outside the scope of this book).

Why we need a GPIO expander

So why do we need a GPIO expander right off the bat, when the Raspberry Pi has so many GPIO pins? This is because PWM pins require computing resources and timers, and many microcontrollers have a limited number of hardware PWM pins. You can emulate PWM pins with software, but the results tend to be on the unreliable side. The Arduino Uno, for example, has eight PWM pins. The Raspberry Pi has only one GPIO pin, and many projects (including the servo and motor projects included later in this book) will require many more than one, and we do not want to use software PWM.

This is why we are using the PCA9685 GPIO expander: it has 16 dedicated PWM pins and provides all the resources to run them. It communicates with the Raspberry Pi using a protocol called I²C (pronounced *eye-squared-see*), the details of which are outside the scope of this book, and are abstracted away in the Johnny-Five component object. See the *Further reading* section if you'd like to learn more about how I²C works.

Wiring up our GPIO expander and RGB LED

First, you'll want to wire the PCA9685 breakout to your PI: GND to GND, VCC to 5V, SDA to SDA, and SCL to SCL. Next, wire a second GND pin from the cobbler to one of the ground lines on the side of the breadboard. Next, our LED: the long leg wires to ground. The leg by itself on one side of the long leg is the red channel; wire that to the 0 column PWM row pin on the PCA9685 board. Green and blue are on the other side; wire them to PWM-1 and PWM-2, respectively. Once this is all done, your project should look similar to the following:

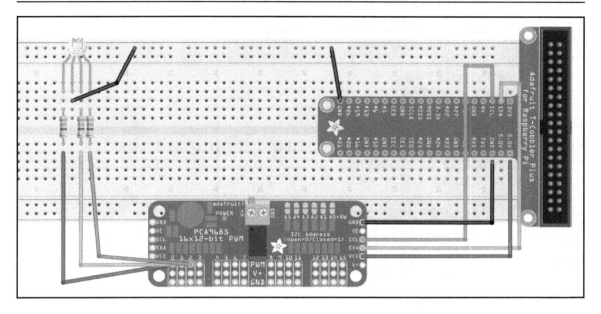

The finished wiring for this chapter's project

Bringing in other node packages

Node.js prides itself on creating small, bordering on tiny, packages, and has the excellent npm package manager (and others) to help manage those packages. Because the Raspberry Pi runs a full version of Node.js, we can leverage that to our advantage and bring in other packages in order to build more interesting projects.

Project – building a rainbow

Can you remember the RGB code for orange off the top of your head? I can't. It's easier to remember to convert the color systems we do know into RGB (especially names such as red, orange, and cornflower blue). But instead of building a function to convert it for us, we'll leverage what I call Stilwell's law: *if you've thought of it, it's probably on npm already*. True to form, the color module is going to help us out.

Using the color npm module

In order to use the `color` npm module, first we will install it. In your SSH session, in your `project` folder, run the following code:

```
npm i --save color
```

This will also save the color package to your `package.json` for code portability purposes. The module exports a function, which we will use to convert color strings like `red` or `#FF0000` to an array of integers representing `red`, `green`, and `blue`. We will use these values to set our RGB LED. This is shown in the following example:

```
const Color = require('color')

let ledColor = Color('orange')
let ledRed = ledColor.red()
let ledGreen = ledColor.green()
let ledBlue = ledColor.blue()
```

We'll use this to help set the color of our RGB LED in our Johnny-Five program.

Getting our Johnny-Five code started

Let's pull together what we've learned about the `Led.RGB` object and the color npm module to pull together a basic code project that we will call `rgb-led-rainbow.js`:

```
const Raspi = require('raspi-io')
const five = require('johnny-five')
const color = require('color')

const board = new five.Board({
 io: new Raspi()
})

board.on('ready', () => {
  let rgbLED = new five.Led.RGB({
  controller: "PCA9685",
  pins: {
    red: 0,
    green: 1,
    blue: 2
  }
  });

  let colors = ['red', 'orange', 'yellow', 'green', 'blue',
```

```
'rebeccapurple']
  let colorIndex = 0
  let currentColor

  setTimeout(() => {
    currentColor = color(colors[colorIndex])
    rgbLED.color([currentColor.red(), currentColor.green(),
currentColor.blue()])
    colorIndex++
    if(colorIndex >= colors.length) {
      colorIndex = 0
    }
  }, 1000)
})
```

This code cycles through the colors in the `colors` array and, once per second, sets the RGB LED's color and moves forward, generating a rainbow.

The REPL – a powerful tool in Johnny-Five

Debugging our LED can be tricky. Without rewiring things, how can we tell if our green and blue channels are flipped, or if the red is far brighter than the other channels? One tool that is very helpful for debugging Johnny-Five projects is the **Read–Eval–Print Loop (REPL)**.

How does the REPL work?

If you have worked with Node.js, Python, or a few other interpreted languages in the past, the REPL may not be new to you. It allows you to write statements into the CLI at runtime to generate results straight from the language engine. This can be very helpful when debugging code, as you can get a glimpse into and modify the state of code at runtime. This is also true in Johnny-Five: the REPL allows us to insert Johnny-Five objects, so we can look at manipulating them at runtime. We're going to use this to play with our RGB LED and control it from the command line.

Adding our RGB LED to the REPL

Take a look at the Johnny-Five documentation for the REPL; it's in the `Board` component section of the API. What matters to us is `this.repl.inject()`, which takes an object and makes any property of that object accessible from the CLI. Let's modify our code to make use of the REPL by making the `rainbow` function check for a Boolean before setting the LED, and adding that Boolean and the RGB LED component object to the CLI:

```
const Raspi = require('raspi-io')
const five = require('johnny-five')
const color = require('color')

const board = new five.Board({
 io: new Raspi()
})

board.on('ready', () => {
  let rgbLED = new five.Led.RGB({
  controller: "PCA9685",
  pins: {
    red: 0,
    green: 1,
    blue: 2
  }
  });

  let colors = ['red', 'orange', 'yellow', 'green', 'blue',
'rebeccapurple']
  let colorIndex = 0
  let currentColor
  let rainbowCycle = true

  setTInterval(() => {
    if(rainbowCycle) {
      currentColor = color(colors[colorIndex])
      rgbLED.color([currentColor.red(), currentColor.green(),
currentColor.blue()])
      colorIndex++
      if(colorIndex >= colors.length) {
        colorIndex = 0
      }
    }
  }, 1000)

  this.repl.inject({
    rainbowCycle,
    rgbLED,
```

```
    color
  })
})
```

Now, we have access to the LED and the Boolean that controls the rainbow cycle from the command-line REPL supplied to us by Johnny-Five when we run this code on our Raspberry Pi.

Controlling our LED from the command-line interface

Move the code over to your Raspberry Pi, and in your SSH session, navigate to your `project` folder using `cd` and run your project (be sure to use `sudo`!):

```
sudo node rgb-led-repl.js
```

Then, you can manipulate the RGB LED and the color library to change the light's color. Here are a few things to try:

```
>> rainbowCycle = false // this stops the rainbow color cycle
>> rgbLED.off() // turns the RGB LED off
>> rgbLED.color(color('rebeccapurple').rgb().array()) // sets the LED
purple!
```

Summary

In this chapter, you created your first interactive project with the Raspberry Pi and Johnny-Five! We started by exploring the LED and LED.RGB APIs, then explored the power that running in Node.js gives us by allowing us to use npm modules, and then we brought it all together with the REPL!

Questions

1. What does PWM stand for, and what does it accomplish with LEDs?
2. Does the Raspberry Pi have any PWM-capable pins? How many?
3. Why do we need a GPIO expander board to control our RGB LED?
4. How many colors would our RGB LED be able to show without PWM?
5. What protocol does our GPIO expander use to communicate with the Raspberry Pi?
6. What does the color module do for us?
7. How does the REPL help with debugging? What makes it so powerful?

Further reading

- **More reading on PWM**: https://learn.sparkfun.com/tutorials/pulse-width-modulation
- **More reading on I²C**: https://learn.sparkfun.com/tutorials/i2c

4
Bringing in Input with Buttons

We've now explored digital and PWM output in Johnny-Five, but that's only half of the story. There is so much you can do with input devices in robotics projects, allowing either user input or observations of the world surrounding your projects to affect the outputs.

We're going to start with a user-input device—buttons. We're also going to talk about how the Raspberry Pi handles digital inputs, and build buttons into our previous project that allows users to stop the rainbow color cycle, and advance the color themselves.

The following topics will be covered in this chapter:

- Using inputs in robotics projects
- The Johnny-Five sensor and button objects
- Wiring up buttons
- Adding buttons to our RGB LED project

Technical requirements

You'll need your Pi, with the RGB LED from the Chapter 3, *Building Interactive Projects with RGB LED* project wired up, along with the GPIO expander board.

Using inputs in robotics projects

You can do a lot with output devices in robotics projects, but the possibilities become endless when you add inputs. Whether they are user-controlled inputs, such as buttons and potentiometers, or environmental sensors that measure things such as ambient light or air quality, input devices can add a new dimension to any robotics project.

Digital versus analog input

Much like with digital and PWM output, there are two types of input devices: digital and analog. Digital inputs are either on or off: buttons are a prime example of this. Analog inputs give a different level of voltage of signal depending on what they are sensing; a photoresistor, for example, puts out higher voltage signals when the ambient light is high, and lower when it is darker.

In order to read data from analog devices, you'll need a pin that can accept an analog input. But as we saw in the last chapter, all of the GPIO pins on the Raspberry Pi are digital. Luckily, there are ways to get around this limitation.

How to handle analog input with the Raspberry Pi

There are two ways to go about obtaining analog sensor data on the Raspberry Pi: adding a GPIO expander that has analog pins, or using sensors that make use of digital signaling to communicate analog data.

Analog GPIO expanders

These boards act almost exactly like the GPIO expansion board we used in `Chapter 3`, *Building Interactive Projects with RGB LED*, except instead of adding PWM output pins, they add analog input pins. These boards also usually utilize an I^2C interface to communicate with the Raspberry Pi. However, I usually find these boards unnecessary, because many sensors that collect more than one channel of data (such as an accelerometer) already utilize I^2C or other digital interfaces, and the few sensors that collect one channel of data can be found with these digital interfaces on board.

Using input devices with digital interfaces

This is the way we'll go in our projects. Devices like these use protocols such as UART, SPI, and I^2C that allow devices that only have digital GPIO to receive analog data. In the materials for each project, the devices included will not require analog input pins.

How Johnny-Five handles input

So we've gotten a glimpse of the way Johnny-Five uses events via the board `ready` event. If you've ever programmed with C and Arduino, you may be familiar with the event loop style of program—a loop runs forever and checks the state of the input devices, then responds accordingly. You may also be aware of interrupt-driven programming, where a change in a hardware pin causes the code to jump to a specific function.

Johnny-Five code is closer to the interrupt style; events drive nearly all Johnny-Five projects. This has several benefits; you can keep your code organized by event type, and make sure each piece of functionality fires only when it needs to, without having to deal with programming your own hardware interrupt routines.

When a Johnny-Five project receives input from a sensor or device, it fires a `data` event. But what if you only want to run a function when the environment changes? The `change` event is for you. We'll look more at the exact event types and when they fire in a later section, but for now keep in mind that events are how you'll capture the data of your sensors and input devices.

The structure of a typical Johnny-Five project

A Johnny-Five project consists of a few key sections and building blocks that make it really easy to read through an example. Let's go through an example here to see more.

The beginning – including libraries and creating our board object

This section sets the stage for us by bringing in the Johnny-Five. The following code snippet tells it we're using a Raspberry Pi, and constructs the appropriate board objects. If you're using other npm modules, like the `color` module we used in Chapter 3, *Building Interactive Projects with RGB LED*, you'd use `require` to bring them in here as well, as shown:

```
const five = require("johnny-five")
const Raspi = require("raspi-io")

let board = new five.board({
 io: new Raspi()
})
```

The board ready event handler

Everything else that we do in a Johnny-Five project, besides the header, goes inside this event handler. This handler, as shown here, means our board is ready to read and write to GPIO pins, and any code run outside this event handler that manipulates GPIO is not guaranteed to work and may cause strange behavior:

```
board.on("ready", function() {
  // Everything else goes in here!
});
```

Constructing our component objects

The first thing I do inside the board ready event handler is set up Johnny-Five objects for all of the components of my projects. It's easier to wire up a project from the code if all of the component types and pins are in the same place in the code:

```
// remember, this goes inside the board ready event handler!
let LED = new five.LED('P1-7')
let button = new.five.Button('P1-8')
```

Input event handlers and output device manipulation

This is where the fun happens, we wait for input and manipulate outputs accordingly! This will watch for a button connected to P1-8 to be pressed, then turn on an LED. But how would we turn the LED off when the button is released? For that, we're going to take another look at the Johnny-Five documentation:

```
// We'll go over more about the Button in the next section!
// This is still inside the board ready event handler!
button.on('press', () => {
  led.on()
})
```

The Johnny-Five button object

Before we program our button project, let's take a good look at the Johnny-Five button object, so we know what events to look for, and what information the constructor wants from us.

The button object

When we look at the button's parameters section, there is only one required parameter, pin. So we'll need to remember what pin to which we hook the signal from the button, but other than that, the defaults will serve us nicely:

- `invert`: Defaults to false, and inverts the up and down values. We'd like to keep this false, as we're wiring the button to not require inversion.
- `isPullup`: Tells boards with pull-up resistors tied to their GPIO pins to initialize this button with the pull-up enabled. We're going to wire our own resistor, so this can stay the default false.
- `isPulldown`: Similar to `isPullup`, but with pull-down resistors. Leave this false as we are wiring our own pull-down resistor.
- `holdtime`: This is the number of milliseconds a button must be held down before the hold event is fired. The default of 500 milliseconds will do fine for us here.

There's also a special section called `collection`, which details how you can control several buttons with the same object. It's an interesting design, and while we won't explore it for our two-button project, a good bonus project would be to refactor it to use the buttons `collection` object.

Button events

There are three events that the button object uses, and each of them can be used on a single instance of a button object:

- `press`, `down`: These are the same event, and they fire when the button has been pressed
- `release`, `up`: These events fire when the button is released
- `hold`: This event fires when the button is held down for longer than the threshold set in the `holdtime` parameter in the constructor

If you've dealt with hardware before, you might be worried about button event `noise`; multiple events firing on one button press or release, release events when the button hasn't been pressed, and so on. Johnny-Five has baked debouncing into the button object, so there's no need to worry about noisy buttons!

Wiring up buttons

We're going to add buttons to the project from the previous chapter, to allow users to change the way the RGB LED works by pressing buttons. When you look at a button, you see four prongs. While there are four prongs, there are only two input/outputs to a button—one where electricity goes in, and one where it goes out when the button is pressed. This is because a button essentially controls the flow of electricity. When the button is not pressed, the contacts are not connected and electricity cannot flow, and when pressed, a conductor bridges the two sides and electricity flows. This is how we will use the button as an input device: a high signal means the button is pressed, a low signal means it isn't.

Putting a button on a breadboard

Take your button and observe the metal prongs on the bottom. Usually, the two pairs curve towards the inside of the button. There is one in and out on each side, and each pair with the same curve has one of each. Keep this in mind when placing it on the breadboard, shown in the following image:

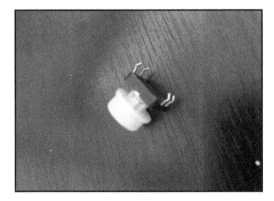

When you're placing a button on the breadboard, you'll want to make sure that the button crosses the trough in the middle of the breadboard in order to prevent a short.

Once you've placed the button into the breadboard, make sure it's well-seated, and that none of the prongs have curled up into themselves instead of going into the breadboard socket. If one has, use needle-nose pliers to straighten it.

Now that you've got your button on the breadboard, it's time to wire it to your Pi.

Using a pull-down resistor

So the question is, how are we going to get three wires into a device with two leads? We're going to use what's called a pull-down resistor to tie an input to the side of the button not connected to power. When the button is pressed, electricity will flow through the resistor into the signal wire, and we'll use a digital input pin on the Pi to detect that as a button press.

To do this, wire one side of your button to a 5V power pin on the cobbler. On the other side, place a 10K ohm resistor that bridges to another row of the breadboard, and in that row place a wire to bridge to a GPIO pin on the cobbler. Then, in the second row of the button, below the resistor, place a wire bridging to a GND pin on the cobbler.

The resistor prevents the Pi from shorting when the button is pressed, which will cause your Pi to temporarily cease functioning and, if left for too long, will cause irreparable damage.

A short circuit, or short, is when the power and ground of a circuit are connected without a load (like our resistor) in between, which causes a lot of issues. To learn more, check out the *Further reading* section of this chapter, or any introduction to an electronics book.

Adding buttons to our RGB LED project

Now that we know how buttons work and how to wire one up to the Pi, let's add two buttons to our RGB LED project.

Wiring everything up

Before we wire up our buttons, we're going to need to do some housekeeping on our current wiring setup.

Using the power and ground side rails

From here on, we'll be needing more access to power and GND pins, and we don't want a ton of really long wires criss-crossing our projects. So the first thing we'll do is a little hardware refactoring.

1. Take the RGB LED ground off the cobbler row.

2. Take the VCC and GND from the GPIO expander off the 5V and GND cobbler rows.

3. Place a wire between the 5V row of the cobbler and the outer long row (if there's one marked red and one blue, use red).

4. Place a wire between a GND pin of the cobbler and the other outer row.

5. Plug the RGB LED ground into the side rail you linked to the GND on the cobbler.

6. Plug the GND from the GPIO expander into the side rail linked to GND on the cobbler, and the VCC into the side rail linked to 5V on the cobbler, as shown in the following diagram:

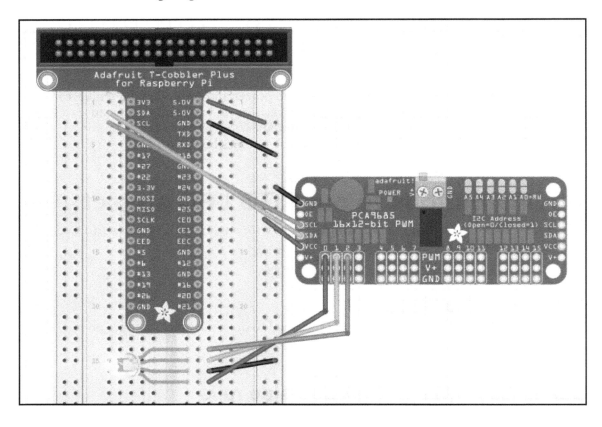

Wiring up the buttons

Now that we've sorted out our power and ground rails, let's place buttons. For both of the two buttons:

1. Place the button on the breadboard as outlined in the last section, bridging the gap in the center of the breadboard.
2. Wire one side of the button to the side rail connected to the 5V pin on the cobbler.
3. Place a 10K ohm resistor on the other side, bridging to an empty rail.
4. Wire the side of the button with the resistor to the side rail linked to GND on the cobbler. Make sure the resistor is in between the button and the link to ground!
5. Wire the other end of the resistors to a pin on the cobbler; use #5 for button 1 and #6 for button 2, as shown in the next diagram:

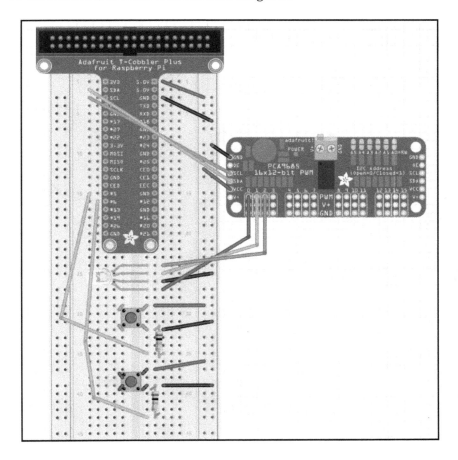

And now you're ready to write some code!

Button 1 – stop and start rainbow

Before we set up our buttons, we're going to refactor our rainbow-cycle program a bit to accommodate the new functionality of the buttons.

Refactoring the rainbow cycle

We're going to refactor the rainbow cycle to do the following:

1. Look at a scoped variable to see if the color should keep changing on a timed interval or not (for the stop and start button)
2. Break the code that changes the RGB LED to the next color into its own function (for the next color button)

Let's take a look at the refactor:

```
board.on('ready', () => {
  let rgbLED = new five.Led.RGB({
    controller: "PCA9685",
    pins: {
      red: 0,
      green: 1,
      blue: 2
    }
  })
  let colorCycle = true

  setInterval(() => {
    if( colorCycle ) {
      colorChange()
    }
  }, 1000)

  function colorChange() {
    console.log(currentColor)
    currentColor = color(colors[colorIndex])
    rgbLED.color([currentColor.red(), currentColor.green(),
  currentColor.blue()])
    colorIndex++
    if(colorIndex >= colors.length) {
      colorIndex = 0
    }
```

```
    }
})
```

We're going to make `button1` stop and start the cycle through rainbow colors. To do this, we'll need to:

1. Construct a button object to represent our button in the code
2. Watch for the `press` event from the Johnny-Five button object API
3. Add a variable called `cycleOn` that can be set to true or false, and have the loop that changes the color use to either change the color or not
4. We're also going to pull the logic for changing the color out in preparation for our next button

Let's add it to the beginning of our board ready handler:

```
let button1 = new five.Button('P1-29')
button1.on('press', () => {
    colorCycle = !colorCycle
})
```

Load this on your Pi, run it with `sudo node rainbow-pause-button.js`, and see what happens when you press the button a few times!

Button 2 – next color

Now we'll add a second button and press handler to make the second advance the color when it is pressed:

```
let button2 = new five.Button('P1-31')

button2.on('press', () => {
    colorChange()
})
```

Now, when you press the second button, the color of the LED will advance to the next color in the array.

Summary

This chapter brought together user inputs and output—an RGB LED. We learned how to use input events in Johnny-Five to manipulate output devices, which is the core of most Johnny-Five projects, and learned how to use multiple inputs (buttons) to achieve different effects.

Questions

1. What events are available to the Johnny-Five button object?
2. Can the Raspberry Pi use analog input devices?
3. How will we use sensors with the Pi?
4. Why are there no events for the RGB.LED object?

Further reading

- **More about analog input pins**: https://learn.sparkfun.com/tutorials/analog-to-digital-conversion
- **More about pull-up resistors**: https://learn.sparkfun.com/tutorials/pull-up-resistors

Using a Light Sensor to Create a Night-Light

In this chapter, we will look at the ways we can still use analog sensors with Johnny-Five and the Raspberry Pi, even without the Pi having built-in analog input pins. We'll use that knowledge to build a night-light that turns on and off an LED based on the ambient light in the room.

The following topics will be covered in this chapter:

- Using an analog sensor with the Pi
- The ambient light sensor
- Creating our night-light

Technical requirements

For this project, you will need a regular LED of any color, and a TSL2561 light sensor, available on Adafruit (https://www.adafruit.com/product/439) and through many other providers.

The code for this chapter is available at https://github.com/ PacktPublishing/Hands-On-Robotics-with-JavaScript/tree/master/ Chapter05.

Using an analog sensor with the Pi

We talked about the lack of multiple PWM output pins on the Pi in `Chapter 3`, *Building Interactive Projects with RGB LED*, but an issue we haven't entirely addressed yet is with inputs. Digital inputs, such as buttons and switches, anything that is either on or off, are easy with the Pi, any digital output pin can also be used as a digital input pin. But what about things that require more than two states, such as sensors that detect light, temperature, moisture, distances, or anything else we'd like to measure in quantity? The answer lies in using specialized communication protocols developed over the years that allow digital pins to communicate analog information.

Finding the right sensors for your Pi project

When you're looking at sensors for a Raspberry Pi project, you need to be sure that any analog sensor you use has a digital interface. The two most common are I²C and SPI, and we'll talk about how to tell which your sensor has (or hasn't!) and whether that device can be used with Johnny-Five.

I²C devices

I²C devices require two more pins to work along with power and a ground pin—an SDA (data) and SCL (clock) pin. The details of how these are used is beyond the scope of this book (see the *Further reading* section for more information), but do know that you can hook multiple devices to the same SDA and SCL pins, so long as the devices have different I²C addresses. The address is a two-digit hex number, that is easy to find for nearly all I²C devices, and in some cases the address can be configured physically on the device.

For this chapter's project, we will be using the TSL2561, which can have two different addresses configured to it. We'll stick with the default, 0x39 (on the Adafruit model) for now.

SPI

SPI devices get tricky quickly, you need five pins: power, ground, microcontroller to sensor data (MOSI), sensor to microcontroller data (MISO), and a chip select line. While multiple devices on a set of SPI pins can share MISO and MOSI pins, they each need their own chip select pin, so the microcontroller can signal the device it wishes to communicate with.

How to determine if your sensor will work with Johnny-Five

The best way to see if there already are drivers for the sensor you are eyeing in Johnny-Five is to check the documentation at the Johnny-Five website. Find the sensor type, and find out what chip the sensor is using (for example, our light sensor uses the TSL2561). Then, on the API page for the sensor, at the top, is nearly always a list of supported controllers and chips. If the chip number on your sensor matches one in that list, it is already compatible with Johnny-Five: just remember that even though analog sensors are compatible with Johnny-Five, they will not work with the Pi, because it has no analog input pins of its own.

As an example, here's the list of supported light-sensor controllers and chips, and you can see the **TSL2561** in the list, so we're good to start building out our project:

Supported Light sensors:

- Photoreistors
  - Sparkfun
    - Mini Photocell
- TSL2561
  - Adafruit
  - Sparkfun
- EV3 Color & Light Sensor
  - Lego
- NXT Color Sensor
  - Lego

The ambient light sensor

To get started with our night-light project, we'll start by wiring up our TSL2561 I²C light sensor and making sure we get good data reads by having it print out to the command line.

Wiring up the sensor

In order to wire up our light sensor, we'll need to know which are the SDA and SCL pins of the Pi. For the Pi 3 and 3 B+, SDA is P1-P3 and SCL is P1-P5; these are also usually labelled on the cobblers as **SDA** and **SCL**. In order to get the sensor working, we'll need the power pin; this sensor is not 5V tolerant, so we'll need to use a 3.3V power pin. We can attach GND on the sensor to any ground pin.

The SDA and SCL pins on the sensor need to be connected to the SDA and SCL pins on the Pi, respectively. In the end, your light sensor should be wired up like the following diagram:

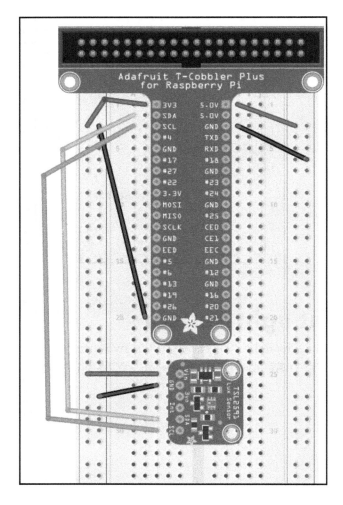

Now that we've wired up our sensor, it's time to figure out how to print that data using Johnny-Five and other Node.js modules so we can make sure it's up and running.

Writing a program to get readings and print them to the command line

Sensor object events in Johnny-Five are different from button events, because, well, sensors are different to buttons! Let's take a look at the differences and how to get the data we need from our light sensor.

The Johnny-Five sensor events

The two main events we'll see from sensors are data and change. The only real difference is in the name: data events are fired every time data is retrieved, while change is fired when the data changes. I tend to use change when building sensor-based projects unless I'm distinctly logging data over time.

You can configure the time between data collection in the construction of the sensor object, as well as the threshold that the change in data must pass in order to fire the change event.

Handling sensor data in the event handler

When you receive data from a sensor, it will be attached to the JavaScript this object, so when you create a callback for the event handler, do not use the arrow syntax, as you will lose the bindings Johnny-Five places on the this object in JavaScript.

Here's an example of a generic data handler for a change event on a Johnny-Five sensor:

```
let mySensor = new five.Sensor('PIN_NUMBER')

mySensor.on('change', function() {
  console.log(this.value) // logs a value between 0-255 to the console
})
```

Now that we've established how we'll get the data, let's talk about what the data will look like and how we can manipulate it.

Using and formatting Johnny-Five sensor data

There are many ways to receive the data sent from a sensor in Johnny-Five, as you can see by the documentation shown in the following screenshot:

Shape

Property Name	Description	Read Only
id	A user definable id value. Defaults to a generated uid	No
pin	The pin address that the Sensor is attached to	No
threshold	The change threshold (+/- value). Defaults to 1	No
boolean	ADC value scaled to a boolean.	Yes
raw	ADC value (0-1023).	Yes
analog	ADC reading *scaled* to 8 bit values (0-255).	Yes
constrained	ADC reading *constrained* to 8 bit values (0-255).	Yes
value	ADC reading, scaled.	Yes
freq	The rate in milliseconds to emit the data event. Disables the event if set to null. (>= v0.9.12)	No

Boolean, raw, analog, constrained, and value can leave you with a lot to process. What each one means is shown in the preceding diagram, however take note that there is a good reason the default value is the same as analog: a scaled reading between 0 and 255. It has a lot to do with the variety of sensors available, the varying granularities of data, and using scaling to make sure you only have to keep one number range in mind, regardless of how many sensors you are using.

Using .scaleTo() and .fscaleTo() to fine-tune measurements

If you'd like to impose an arbitrary scale on your sensor (say $0 - 100$ for percentage), you have some options built into the Johnny-Five API: `.scaleTo()` and `.fscaleTo()`. These will scale the raw value from the sensor to match the min and max values you pass in:

```
sensor.on('change', function(){
  // this.value will reflect a scaling from 0-1023 to 0-100
  console.log(this.scaleTo([0, 100])); // prints an integer
  console.log(this.fscaleTo([0, 100])); // prints a float
})
```

Now that we know how to handle the data, let's start on our night-light by creating code to print the light-sensor values to the command line. This will also allow us to tweak the change threshold setting and determine what value of the light sensor we should use as an indicator to turn our LED off and on.

Printing sensor data to the command line

To print data from our sensor to the command line, we'll use the code in `print-light-sensor.js`:

```
const five = require('johnny-five')
const RaspiIO = require('raspi-io')

let board = new five.Board({
  io: new RaspiIO()
})

board.on('ready', () => {
  let lightSensor = new five.Light({
    controller: 'TSL2561'
  })
  lightSensor.on('change', function() {
    console.log(this.value)
  })
})
```

Your output on run should look something like this, with the numbers varying when you cover or shine light onto the sensor:

```
Code/misc/barcli-ex is ● v1.0.0 via ● v10.5.0
➜ node index.js
Light Sensor:   254
Light Sensor:   200
Light Sensor:   7
Light Sensor:   88
Light Sensor:   81
Light Sensor:   80
Light Sensor:   10
Light Sensor:   78
Light Sensor:   195
Light Sensor:   162
Light Sensor:   254
Light Sensor:   151
Light Sensor:   127
Light Sensor:   132
Light Sensor:   175
Light Sensor:   233
```

This is nice, but a little hard to comprehend. What we'll do next is add in the npm module barcli to show a nice bar graph that allows us to comprehend in real time the data we're seeing.

Using barcli to make the data easier to see

That data stream can be hard to process! Let's take a look at leveraging the power of Node.js to make this easier to see.

In your project folder, run:

```
npm i --save barcli
```

To install barcli, a library that creates bar graphs in the Terminal.

Reading the barcli documentation (see *Further reading*), we'll need to import barcli, construct a barcli object with the settings we need, then tell when to update and with what data.

Importing barcli and constructing our barcli graph

To import `barcli`, at the top of your `print-light-sensor.js` file, following the other `require()` statements, add:

```
const Barcli = require('barcli')
```

Then, in the `board.on('ready')` handler, we'll add the bar graph constructor:

```
let lightGraph = new Barcli({
  label: 'Light Sensor',
  range: [0, 255]
})
```

Getting the bar graph to update

Remove the `console.log()` line from the `lightSensor.on('change')` handler and replace it with:

```
lightGraph.update(this.value)
```

Then you're ready to roll! Move the `project` folder over to the Pi, navigate to the folder in your Pi SSH session, and run:

npm i

To make sure that `barcli` is properly installed on the Pi, run the command:

sudo node light-sensor-barcli.js

You should see a bar graph now, shown as follows, that changes when you shine light on or cover the sensor:

Now, for our night-light project, you'll want to find a value for the light sensor that we will use to turn the LED on and off; `barcli` makes this much easier by making that value much easier to see.

Once you've got the value that works for you (I settled on 25), we're ready to build our night-light.

Creating our night-light

Now that we know our light sensor works, we can add an LED and create our night-light.

Wiring up the LED

Connect the short leg of your LED to a ground rail using a 330K ohm resistor, and wire the long leg to GPIO #5, also known as P1-29:

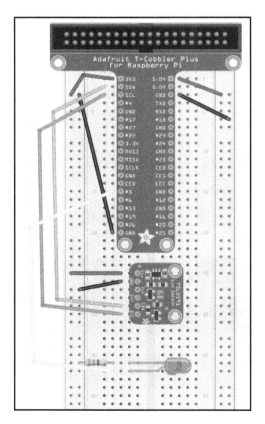

Coding this project

Create a file in the same folder as the other files from this chapter, and copy the contents of `print-light-sensor-readings.js` into it.

In the start of the `board.on('ready')` handler, add a constructor for our LED:

```
let light = new five.Led('P1-29')
```

And in the `lightSensor.on('change')` function, replace the `console.log` statement with the logic that will turn the LED on and off:

```
if(this.value <= 25) {
  light.on()
} else {
  light.off()
}
```

And we're ready to run! Load the folder onto your Pi, navigate to the folder in your Pi's SSH session, and run:

```
sudo node night-light.js
```

When you cover the light sensor with your thumb, the LED should light up, as shown in the following image:

And when you remove your thumb (in a well-lit room), the LED will turn off, as shown in the following image:

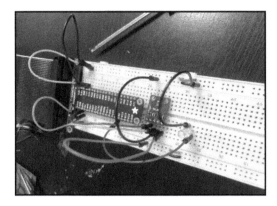

And with that, you have successfully coded and built your night-light!

Summary

In this chapter, we learned about analog sensors and the limitations of the Pi for analog input. We learned about digital interfaces that allow us to collect analog data in Pi projects. We used this knowledge to set up a light sensor, with bar graphs from `barcli` to find a good threshold for an LED to turn on and off. Finally, we used all of this together to build a night light that illuminates in the dark and turns off in the light.

Questions

1. What is an analog input sensor?
2. Why can't analog input sensors directly interface with the Raspberry Pi?
3. Name two digital interfaces we can use with the Pi to collect analog data.
4. What two pins (besides power and ground) do I^{2}C sensors need to operate?
5. Name the events that a sensor object can fire.
6. Why is `barcli` helpful in processing changing sensor data?

Further reading

- **More information on analog inputs**: `https://learn.sparkfun.com/tutorials/analog-to-digital-conversion`
- **More information on SPI**: `https://learn.sparkfun.com/tutorials/serial-peripheral-interface-spi`
- **More information on I²C**: `https://learn.sparkfun.com/tutorials/i2c`
- **More information on using SPI and I²C with the Pi**: `https://learn.sparkfun.com/tutorials/raspberry-pi-spi-and-i2c-tutorial`

6
Using Motors to Move Your Project

We've explored using input to discover the world around our bots, and output to let our bots communicate, but there is another crucial skill any bot should have: the ability to move! In the next few chapters, we'll discuss various ways we get let our bots to move, and discuss how to control that movement. We'll start in this chapter with the simplest movement component: the motor.

The following topics will be covered in this chapter:

- More about motors
- Preparing for a motor-driven project with the Raspberry Pi
- The Johnny-Five motor object
- Troubleshooting your motorized projects
- Project – building a randomized motorized cat toy
- Project – using a gearbox motor and the motors object

Technical requirements

You will need the Adafruit Pi motor hat kit (`https://www.adafruit.com/product/2348`), a small 5V motor, which you can also get from Adafruit (`https://www.adafruit.com/product/711`), or many other suppliers, a 4-AA battery case with wire ends and an on/off switch, available from Adafruit (`https://www.adafruit.com/product/830`) and many other suppliers, 2 *gearbox* or *TT* motors, available from Adafruit (`https://www.adafruit.com/product/3777`) and many other suppliers, and a sticky note (or a piece of paper, scissors, and tape).

Note: If you cannot solder or are uncomfortable soldering, an alternative fully assembled hat can be found on Amazon (`https://www.amazon.com/SB-Motorshield-Raspberry-expansion-ultrasonic/dp/B01MQ2MZDV/ref=sr_1_fkmr1_1?s=electronicsie=UTF8qid=1534705033sr=8-1-fkmr1keywords=raspberry+pi+motor+controller+TB6612`). I will note changes in the code where necessary—anytime this chapter references the L293D hat, that is in reference to this hat.

The code for this chapter is available at `https://github.com/PacktPublishing/Hands-On-Robotics-with-JavaScript/tree/master/Chapter06`.

More about motors

A motor is a component that can rotate a shaft in continuous circles at varying speeds. However, there are many different kinds of motors; let's take a look at a few:

- **DC motor**: This kind of motor is the simplest: it can go in one direction, and the speed varies by the power you give it. These usually only have two wires: one for ground and one for power; we will combine the latter with the motor hat to control the speed. With the correct controller, we can move the motor in both directions.

- **Motors with brakes**: These motors have a third wire to control a brake that can stop the motor without the need to coast to a stop, as with DC motors. These motors are supported by the Johnny-Five library, but will not be covered in this book.

- **Stepper motors**: Stepper motors are used for precise movements, as they move in steps that vary based on the size of the motor. They are bi-directional by design and are great where you need the torque of a motor with precision. We'll talk more about these in the second project in this chapter. Just know that an easy way to spot a stepper motor is 5 wires as opposed to 2 or 3:

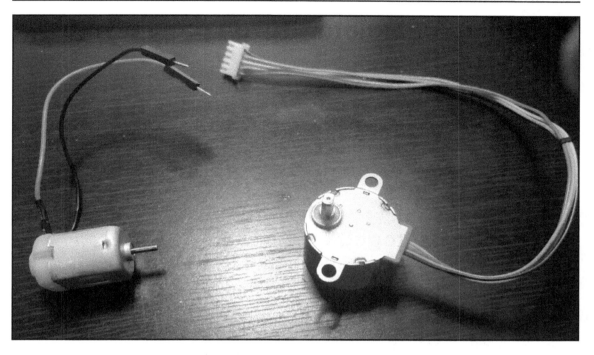

Regular motor on the left, stepper motor on the right

How to control a motor with a microcontroller

You can directly connect DC hobby motors to the PWM pin of a microcontroller to power them, but this is usually inadvisable: motors take up a lot of power and many microcontrollers limit the amount of current out of each pin.

A more advisable solution, which we will be using in this chapter, is to use an external motor controller; these usually contain the circuitry necessary to do more complex movement with your motors (such as allowing them to go backwards), and allow for an external power supply that provides the necessary power to your motors without drawing it from the microcontroller.

Preparing for a motor-driven project with Raspberry Pi

In order to get started with motors using Johnny-Five and the Raspberry Pi, we'll need to add a hat (think Arduino shields, but for the Pi, or add-on boards that stack on the Pi if you're new to electronics) that allows us to:

- Provide external power to the motors
- Control the motors better than the Pi can on its own (especially in the case of the stepper motor)

Putting the hat together

Wire the battery pack and the motor to the hat's screw terminals like so:

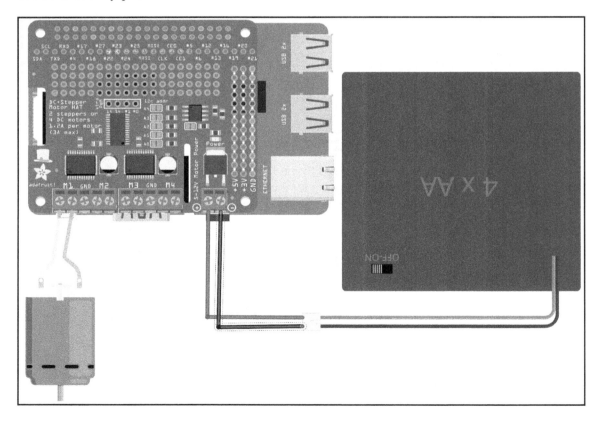

The yellow wire in the diagram should be your ground wire (usually black) and the green should be your power wire (usually red).

Putting the hat on the Pi

Remove all power from the Pi, and make sure the battery pack is switched off. Also, remove the cobbler from the GPIO pins if it is still attached. Then, line up the sockets on the bottom of the hat with the pins on the top of the Pi, in the direction that makes it so the hat is situated over the Pi. Then, gently press down on the hat until it settles. **Don't press hard— you may bend some of the Pi's pins**. When all is said and done, it should look something like this:

And the motor should be plugged into the screw terminals, like so:

Re-apply power to the Pi, and we'll get started coding using the Johnny-Five motor object.

The Johnny-Five motor object

The motor object in Johnny-Five allows us to easily control our motors without having to worry about communicating with the hat via the Pi. Let's code a test setup with the REPL before coding our project, to make sure everything is working.

Create a new `project` folder and, inside it, run the following:

```
npm init -y
```

And, create a file in the folder named `motor-test.js`. Start by requiring in Johnny-Five and Raspi-IO, instantiating your board object, and creating a `board.on('ready')` handler, as we usually do:

```
const Raspi = require('raspi-io')
const five = require('johnny-five')
const board = new five.Board({
  io: new Raspi()
})

board.on('ready', () => {

})
```

Now, we're ready to set up our motor object, keeping in mind that we'll need to configure for our hat.

Constructors for our hat

If you are using the Adafruit hat, your constructor is as follows:

```
let motor = new five.Motor(five.Motor.SHIELD_CONFIGS.ADAFRUIT_V2.M1)
```

And if you're using the L293D hat, your constructor is as follows:

```
let motor = new five.Motor(five.Motor.SHIELD_CONFIGS.ADAFRUIT_V1.M1)
```

Place whichever one applies inside the `board.on('ready')` function.

Functions that move the motor

Referencing the Johnny-Five documentation, there are a few functions that will allow us to move the motor from the command line using the REPL:

```
motor.forward(speed) // speed 0-255, starts the motor forward
motor.stop() // the motor coasts to a stop
motor.start(speed) // resumes the motor moving forward
motor.reverse(speed) // moves the motor backward
```

Now that we know how to control the motor, let's add the REPL functionality to test it

Adding REPL control

At the end of the `board.on('ready')` function, add the following:

```
board.repl.inject({
  motor
})
```

And now we have full control!

Loading and running your motor

Load the project onto the Pi, and navigate to the folder in the Pi ssh session. Then, run the following:

```
npm i --save johnny-five raspi-io
```

Once that's completed, turn on the battery pack and run the following:

```
sudo node motor-test.js
```

Once you see *Board Initialized*, you should try out some of the commands from before:

```
motor.forward(speed) // speed 0-255
motor.stop()
motor.start(speed)
motor.reverse(speed)
```

Hopefully, your motor is happily spinning away!

Troubleshooting your motorized projects

But what if your motor doesn't turn? Here are a few things to check if your motor isn't spinning around:

- Is the battery pack for the motor hat turned on? Don't laugh, I've spent many a minute wondering why it wasn't working only to discover it lacked power. There's a power light on most motor hats that let you know it has power:

The power LED on the Adafruit hat is just above the screw terminals for the external power, and lights up red

- Are your batteries fresh? Motors take up a lot of power, and extended use can wear them down pretty fast.
- As I mentioned in the first chapter: check your wiring. Then, check it again.
- Make sure all of the wires are securely fastened in the correct screw terminals so that a light yank cannot dislodge them.
- Are you using rechargeable batteries? If so, I admire your commitment to reuse, but you're going to want 6 rechargeable running your motor due to differences in voltages between rechargeable and non-rechargeable batteries.

Hopefully, if your motor wasn't spinning before, it is now, and we can build our first project.

Project – cat toy

In this project, we'll add a piece of paper to our motor, and then code some randomness to make it spin back and forth at varying speeds (cats get bored with a predictable toy, after all).

The wiring for this project is the same as the motor test; no need to change anything there.

Putting a piece of paper on the motor shaft

Either roll the sticky end of a long sticky note around the motor shaft, or tape a long strip of paper to it. It should look something like this:

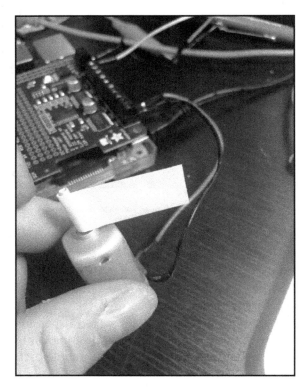

After the relatively simple construction of our toy, let's code some randomness!

Coding the randomness to start/stop the motor

We want the motor to start at a random speed for anywhere from 1-10 seconds, then stop for 1-10 seconds, and repeat. We also want whether it goes forward or backward to be random. I limited the speed to 75—anything faster was too much for my cats!

In your `cat-toy.js` file, get rid of the `board.repl.inject` statement and add the following:

```
startMovement()

function startMovement(){
  let direction = Math.round(Math.random())
  let speed = Math.round(Math.random() * 75)
  let time = Math.round(Math.random() * 10)

  if(direction == 0){
    motor.forward(speed)
  } else {
    motor.reverse(speed)
  }

  setTimeout(stopMovement, time*1000)
}

function stopMovement(){
  let time = Math.round(Math.random() * 10)
  motor.stop()
  setTimeout(startMovement, time * 1000)
}
```

This will randomize the starting and stopping, the speed, and the direction. My cats were at least mildly entertained by it. If you have cats, give it a try!

Load the project onto the Pi, and navigate to the folder in the Pi SSH session. Then, run:

```
sudo node cat-toy.js
```

And watch it go!

We've got one motor going, but if you want to build a bot with wheels, we're gonna need two motors; let's take a look at that concept with our next project.

Project – using two gearbox motors and the motors object

Now that we've explored the motor object, let's dig a little deeper and build a project using two TT motors while exploring the motors object.

If you want to take this a step further, you can get yourself a chassis like this one from Adafruit `https://www.adafruit.com/product/3796` and a pair of wheels like these from Adafruit `https://www.adafruit.com/product/3757` and build yourself a moving 2-wheel robot! Just remember you'll have to either power the Pi with a battery pack (those little USB packs for charging your phone work great) or stay within range of your Pi's power cord. If you go with the latter, I'd secure the power jack into the Pi and be very careful not to let the bot pull too hard on anything. Honestly, I'd really just recommend using a battery if you're going to let your Pi move about on its own.

Wiring up your TT motors

For this diagram, pretend the normal DC motors are our TT motors—yellow will be the ground (usually black) wire, and the green will be the power (normally red) wire.

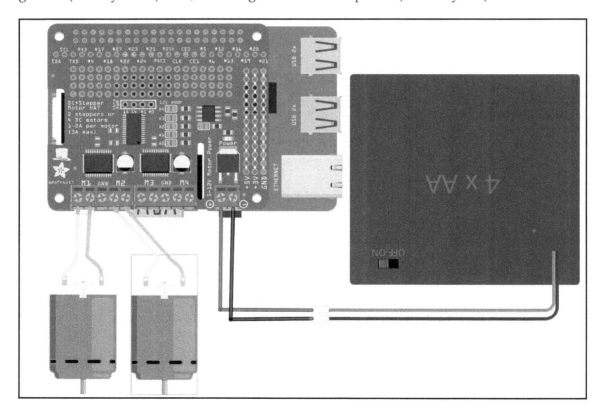

Now it's time to get started coding our motors to perform common wheeled-vehicle movements using Johnny-Five and the motors object.

The motors Johnny-Five object

Create a new file in your project folder called `driver-bot.js`. Start with the usual setup of the Johnny-Five and Raspi-IO libraries, your board object, and your `board.on('ready')` handler:

```
const Raspi = require('raspi-io')
const five = require('johnny-five')
const board = new five.Board({
  io: new Raspi()
})

board.on('ready', () => {

})
```

Next, inside the `board.on('ready')` handler, we'll add the constructors for our two TT motors:

If you are using the Adafruit hat, your constructors are:

```
let leftMotor = new five.Motor(five.Motor.SHIELD_CONFIGS.ADAFRUIT_V2.M1)
let rightMotor = new five.Motor(five.Motor.SHIELD_CONFIGS.ADAFRUIT_V2.M2)
```

And if you're using the L293D hat, your constructors are:

```
let leftMotor = new five.Motor(five.Motor.SHIELD_CONFIGS.ADAFRUIT_V1.M1)
let rightMotor = new five.Motor(five.Motor.SHIELD_CONFIGS.ADAFRUIT_V1.M2)
```

Now that our motors are constructed, we'll create our Motors object by passing it an array containing `leftMotor` and `rightMotor`:

```
let motors = new Five.Motors([leftMotor, rightMotor])
```

Before we start writing our driving functions, let's talk a little about the benefits of the Motors object. The main benefit to having your motors in a Motors object is to maintain control over each individual motor while also being able to control them all at once. For example:

```
leftMotor.forward(255) // left motor full speed ahead!
rightMotor.reverse(255) // right motor full speed reverse!
motors.stop() // both motors coast to a stop
```

The motors object allows you to call any Motor object function and it will perform it on all of the motors at once.

Let's use this knowledge to write some common driving functions that we can use with our motors.

Writing some functions

First things first, we'll want to let our robot go forward. Inside the `board.on('ready')` handler of `driver-bot.js`, add:

```
function goForward(speed) {
  motors.forward(speed)
}
```

Here we see again the benefit of the `motors` object; we don't have to tell the right and left motor to move forward separately.

Let's add another function to let our motors coast to a stop:

```
function stop() {
  motors.stop()
}
```

And another to let our robot go backward:

```
function goBackward(speed) {
  motors.reverse(speed)
}
```

Now that those are done, how about we add some turns? Luckily, the `motors` object still lets us control each motor individually—so turns are no problem:

```
function turnRight(speed) {
  leftMotor.forward(speed)
  rightMotor.stop()
}

function turnLeft(speed) {
  rightMotor.forward(speed)
  leftMotor.stop()
}
```

Finally, let's add the ability to spin left or right in place:

```
function spinRight(speed) {
  leftMotor.forward(speed)
  rightMotor.reverse(speed)
}

function spinLeft(speed) {
  rightMotor.forward(speed)
  leftMotor.reverse(speed)
}
```

Now, our motors have everything they need to drive around! Let's give ourselves REPL access to these methods, the `motors` object, and the `motor` objects:

```
board.repl.inject({
  leftMotor,
  rightMotor,
  motors,
  goForward,
  goBackward,
  stop,
  turnRight,
  turnLeft,
  spinRight,
  spinLeft
})
```

And now we're ready to load up our code and take our motors for a spin (both metaphorically and literally)!

Running our motors project

Load the project onto the Pi, and navigate to the folder in the Pi SSH session. Then, run:

```
sudo node driver-bot.js
```

Once you see *Board Initialized*, feel free to try out our new functions:

```
goForward(100) // start moving both motors forward
stop() // and stop
goBackward(50) // go backward at half the previous speed
stop()
goForward(100)
turnRight(100)
turnLeft(200) // a faster left turn
```

```
spinRight(255) // robots can't get dizzy, maximum fastness
spinLeft(100)
stop()
```

And that's it—you've written all the code you need to drive a two-wheeled robot!

As a bonus project, think of a way you could drive the bot without having to type out the function names each time!

Summary

In this chapter, we learned about the first component that adds movement to our bots: the motor. We learned about the types of motors, and how to interface one with a microcontroller. Then, we wrote code to test our motor with the Pi hat and the REPL, and we built a small randomized cat toy using our knowledge of the Johnny-Five Motor object. Finally, we built a project that allowed us to explore hands-on the abilities of the Motors object and write code to drive a two-wheeled robot.

Questions

1. What is a motor?
2. What's the difference between a motor and a stepper motor?
3. Why should you use external power for motors?
4. Why do we need a hat to control our motor?
5. What are the benefits of the Motors object when using multiple motors?

Using Servos for Measured Movement

7

Now that we've looked at motors, let's look at a more precise way to add to movement to our projects: the servo. We'll dive into how to wire up one servo, then another, and how to code both a single servo and multiple ones.

The following topics will be covered in this chapter:

- Differences between motors and servos
- Getting a servo working with Johnny-Five
- Project – two servos and the REPL
- Project – the continuous servo

Technical requirements

You'll need your Pi, Cobbler, PWM hat, 2 hobby servos, and the AA battery pack from Chapter 6, *Using Motors to Move Your Project*. You'll also need a continuous servo. Finally, you'll need your light sensor. Optional, but helpful, is a Popsicle or other small stick and some tape for turning our servo into a meter.

 The code for this chapter is available at: https://github.com/PacktPublishing/Hands-On-Robotics-with-JavaScript/tree/master/Chapter07.

Differences between motors and servos

The average hobby servo looks like this:

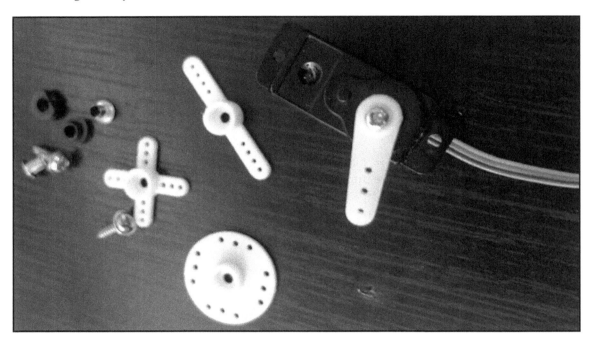

Also pictured are common accessories: some different arms, and mounting screws/washers. The wire ends terminate in a solid socket of three: perfect for attaching to our PWM hat.

The motors we worked with last chapter have some very basic differences, and we should explore them before we start our servo project.

Calculated movements

Unlike the motors we dealt with in the previous chapter, you can make precise and calculated movements with a servo. Name a degree between 0 and 180 on a regular servo, and it will go. Motors (excluding stepper motors, which aren't covered in this book) cannot make these precise movements. So if you're looking to make a wheel go and you don't care about accurate movements, use a motor. When you're looking to make the joint of a limb that needs to move precisely with other joints, time to use a servo.

Regular versus continuous servos

There are two kinds of servos, and they look remarkably alike. The other kind is called a continuous servo. They work very similar to regular servos in Johnny-Five, as you can see in the documentation for *servo.continuous,* which is on the same page as the servo documentation. The main difference is that while a regular servo can only go 180 degrees, a continuous one can go full 360 and continue spinning in the same direction indefinitely. Some favor using these to move wheels in their vehicle projects, and that's just fine!

Powering servos and motors

This is where servos and motors are very similar: while one servo tends to take less power than a motor, many projects make use of more servos than they would motors, and that can mean a strain on your Pi. If you add a lot of servos, you're going to want to provide external power to the PWM hat; many use 3-4 AA battery packs. I recommend using the one we used for motors in `Chapter 6`, *Using Motors to Move Your Project.*

 Powering your servos off of the Pi can have consequences that you wouldn't expect-- if you're getting memory leak issues while running servo code, it can be because you're pulling too much power from the Pi! In general, if you're having weird issues with running servo code, make sure your servos are adequately powered by an external source.

Getting a servo working with Johnny-Five

To get a servo working with Johnny-Five, we'll look at the Johnny-Five servo object, talk about wiring the servo to our PWM hat, and write our first piece of code to get the servo to sweep back and forth.

The Johnny-Five servo object

Looking at the servo page in the API section of the Johnny-Five documentation, we will look first for our constructor. Because we're still using the PCA9685 PWM hat, our constructor will look like this:

```
let servo = new five.Servo({
  controller: "PCA9685",
  pin: 0
});
```

As for moving the servo, there are a few method described in the docs to move the servo. The first move can be to a fixed position:

```
servo.to(degree)
servo.min()
servo.max()
servo.home()
servo.center()
```

Or, another is to sweep back and forth, either as far back and forth as possible, or between a range:

```
servo.sweep() // goes 0-180 and back, then repeats
servo.sweep(minDegree, maxDegree) // goes min to max and back, then repeats
```

You can also stop a moving servo:

```
servo.stop()
```

Now that we have a good grasp on coding our servo in Johnny-Five, let's wire up a servo and test our new knowledge.

Wiring the servo to our PWM hat

Wiring the servo to the PWM hat is relatively simple; you need to line up the 3-pin socket of the servo to the column of pins you want. You also need to attach the power leads of the shield to AA battery pack from Chapter 6, *Using Motors to Move Your Project*.

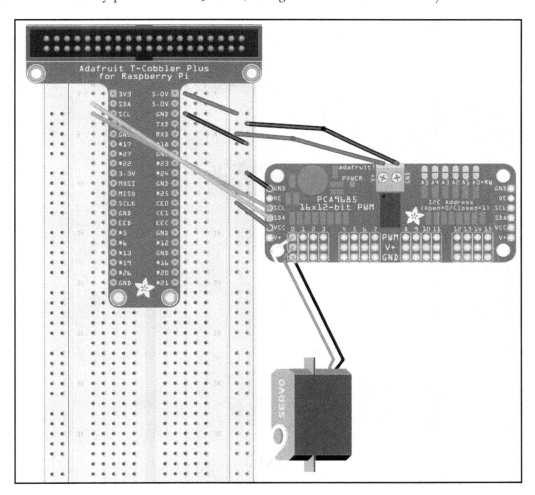

I find that figuring out which pin is ground, then making sure that socket is on the bottom, helps. Ground wires on servos are typically black or brown in color. Then, line up the socket and slide it onto the pins in the first column (pin 0).

We've hooked up our servo and given power to the PWM hat, so let's code up a servo.

Coding your first servo sweep

In a file called `single-servo.js`, we're going to set up our board, the our servo, and when it's ready, tell it to sweep back and forth:

```
const Raspi = require('raspi-io')
const five = require('johnny-five')

const board = new five.Board({
 io: new Raspi()
})

board.on('ready', () => {
 let servo = new five.Servo({
 controller: "PCA9685",
 pin: 0
 })

 servo.sweep()
})
```

Move the file into its own folder on the Pi, navigate to that folder in your Pi SSH session, and run the following:

```
npm init -y
npm i --save johnny-five raspi-io
sudo node single-servo.js
```

You should see the servo move back and forth. Great! Now it's time to control two servos using the Johnny-Five Servos object and the command-line REPL.

Project – two servos and the REPL

Now that we have one servo up and running, we're going to wire up a second one and use the REPL to explore the Johnny-Five Servos object, which is meant to help control several servos at once.

First, let's wire up our second servo.

Wiring up a second servo

Take the second servo, figure out which side is ground, put that one on the bottom, and slide the three-pin socket over the pins in the second column (pin 1):

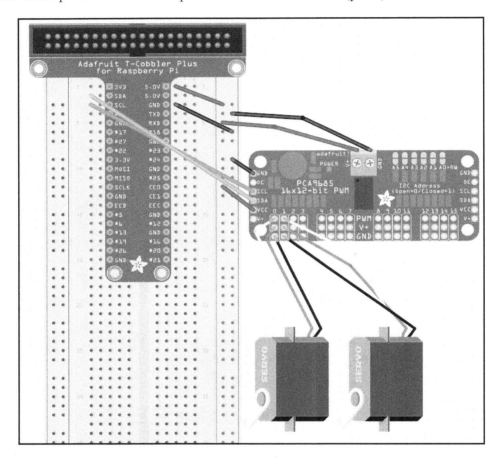

Now that we've wired up a second servo, let's start coding our Johnny-Five servos object!

Using the Johnny-Five servos object

The Johnny-Five servos object is meant to help you group servos in ways that make sense for projects with many servos, such as hexapods with six legs, each containing multiple servos.

You can create a `Servos` object in a few different ways; the way we will use is to pass an array of constructed `servo` objects:

```
let servos = new five.Servos([servoOne, servoTwo])
```

This is where the magic happens—now that our `servo` objects are grouped in a `Servos` object, we can control them both independently and as a group:

```
servoOne.to(0) // Sets servoOne to 0 degrees
servoTwo.to(180) // Sets servoTwo to 180 degrees
servos.center() // Sets both servos to 90 degrees
```

And *all servo object functions are available on the Servos object.*

Let's add this to our code.

Adding the Servos object to our code

In the same folder as `single-servo.js`, create a new file, `servos-repl.js`, and copy the contents of `single-servo.js` into it.

Then, in the `board.on('ready')` handler, rename `servo` to `servoOne` and add a constructor for `servoTwo` on pin 1 of the PWM hat:

```
let servoOne = new five.Servo({
  controller: "PCA9685",
  pin: 0
})

let servoTwo = new five.Servo({
  controller: "PCA9685",
  pin: 1
})
```

Then, we'll construct a `Servos` object containing `servoOne` and `servoTwo`:

```
let servos = new five.Servos([servoOne, servoTwo])
```

Now that our servos are coded, let's add Johnny-Five REPL functionality so we can control our servos from the command line.

Adding in REPL functionality

First, delete the `servo.sweep()` line from `servos-repl.js`, as it'll cause an error now. Instead, place this code, which will allow us to access both of our servos and our servos group object from the command-line, on the Pi:

```
board.repl.inject({
  servoOne,
  servoTwo,
  servos
})
```

Now we're ready to play with our servos!

Playing with our servos on the command line

Load the folder onto the Pi, navigate to the folder in your Pi SSH session, and run the following:

```
sudo node servos-repl.js
```

Once you see `Board Initialized`, the REPL is ready for commands. Try these out for starters:

```
servoOne.to(0)
servoTwo.to(180)
servos.center()
```

Use the Johnny-Five documentation on servo to see what other fun things you can try with your servos from the command line!

Project – light meter with the servo

Let's build a project where our servo servos as a light meter that sweeps between 0 and 180 degrees based on the reading from the light sensor.

Adding in the light sensor

First, we need to wire the light sensor to the board. Remember, I^{2}C devices can share an SDA and SCL pin as long as they have different addresses (which the TSL2591, at 0x29, and the PWM hat, at 0x40, do).

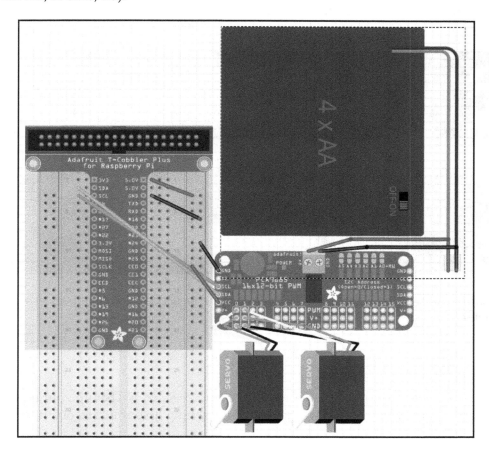

Now that we've wired up the sensor, we'll take on the (optional) task of modifying our servo to look more like a meter.

Making the servo into a meter

Take the servo horn and, with the center of the horn facing away from you, move it as far to the left as possible (0 degrees). *Do not use a lot of force or you'll strip the gears.* Then, use the tape to tape your stick to the servo horn to make it appear longer. then, you can tape it down to a desk or onto a wall, with your meter pointing left. You can see my attempt in the following diagram:

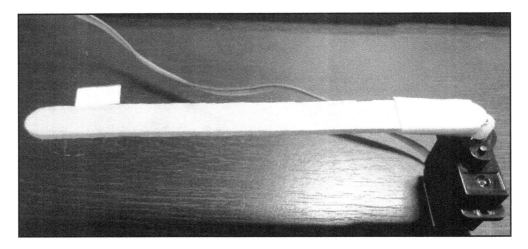

Now that we have our light sensor wired up and our meter cobbled together, let's code our light meter.

Coding the project

Create a new file in your `project` folder called `light-meter.js`. Set up your normal scaffolding: requiring in Johnny-Five and Raspi-IO, setting up our `Board` object, and creating the `board.on('ready')` handler:

```
const Raspi = require('raspi-io')
const five = require('johnny-five')

const board = new five.Board({
  io: new Raspi()
})

board.on('ready', () => {
})
```

Inside the `board.on('ready')` handler, construct your `Servo` and `Light` objects:

```
let servo = new five.Servo({
  controller: "PCA9685",
  pin: 0
})

let lightSensor = new five.Light({
  controller: 'TLS2561'
})
```

Then, we need to build a `lightSensor.on('change')` handler. We're going to use the `Sensor.scaleTo([min, max])` to scale the 0-255 input of the light sensor to the 0-180 output of the `servo`:

```
lightSensor.on('change', function(){
  servo.to(this.scaleTo([0, 180]))
})
```

And that's it! Let's get it on the Pi and see it at work.

Running and using our light meter

Load your `project` folder onto the Pi, navigate to the folder in your Pi SSH session, and run the following command:

```
sudo node light-sensor.js
```

Then, cover the light sensor, or shine a light onto it—the servo should move back and forth accordingly.

Now that we've explored the servo as a means of communicating data as well as creating motion, let's take a look at the continuous servo.

Project – the continuous servo

Continuous servos are a little like a motor combined with a regular servo: you lose the ability to go to specific degrees like a regular servo, but you can stop instantly instead of coasting to a stop like many motors. You can tell a continuous servo to move clockwise or counterclockwise at different speeds, and you can tell it to stop.

Let's wire up a continuous servo and play with its abilities via the Johnny-Five REPL.

Wiring up the servo

The only difference in the wiring is continuous servos look different from regular servos in that nearly all have a disc instead of a horn. And most have red, white, and black power, signal, and ground wires, respectively.

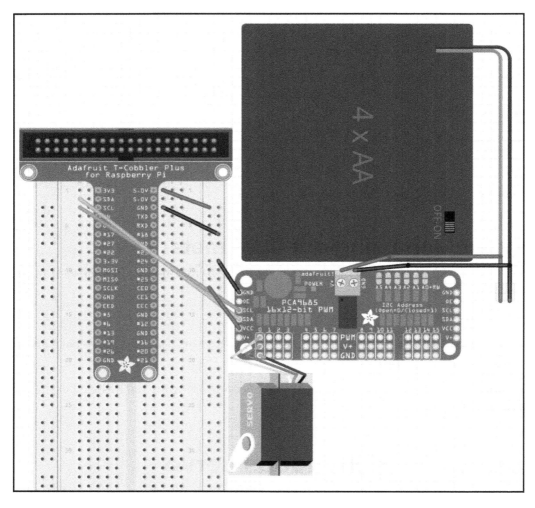

(Fritzing didn't have a continuous servo object, so we'll have to make due.)

Continuous servo constructor and methods

The constructor for the continuous servo is reminiscent of the RGB LED constructor, in that it is a property of the `Servo` object. Otherwise, it looks very similar to the `Servo` constructor with our PWM hat:

```
let continuousServo = new five.Servo.Continuous({
  controller: "PCA9685",
  pin: 0
})
```

There are three methods we can use with the continuous servo:

```
Servo.Continuous.cw([speed of 0-1]) // turns the servo clockwise
Servo.Continuous.ccw([speed of 0-1]) // turns the servo counter-clockwise
Servo.Continuous.stop() // stops the servo
```

Now that we know how to use it and have it wired up, let's write a quick program that lets us play with our continuous servo in the REPL.

Using the REPL with the continuous servo

Create a file in your `project` folder called `continuous-servo-repl.js`, and start with your usual setup:

```
const Raspi = require('raspi-io')
const five = require('johnny-five')

const board = new five.Board({
  io: new Raspi()
})

board.on('ready', () => {
})
```

Then, in our `board.on('ready')` handler, construct the `Servo.Continuous` object:

```
let continuousServo = new five.Servo.Continuous({
  controller: "PCA9685",
  pin: 0
})
```

Finally, after the constructor, inject the continuous servo into the REPL so we can use it:

```
board.repl.inject({
  continuousServo
})
```

Now we can load it up and try it out!

Playing with the continuous servo in the REPL

Load your `project` folder onto the Pi, navigate to the folder in your Pi SSH session, and run the following command:

```
sudo node continuous-servo-repl.js
```

Once you see `Board Initialized`, you can try controlling the continuous servo by entering commands into the Johnny-Five REPL:

```
continuousServo.cw(1) // see how fast it can go!
continuousServo.ccw(.5) // it changes directions near-instantaneously!
continuousServo.stop() // and stops very quickly, too!
```

It is because of the ability to change direction and stop quickly that some prefer continuous servos to motors for wheeled robots—it's always great to have options.

Summary

In this chapter, we learned how servos differ from motors, and the difference between regular and continuous servos. We also learned about the constructor for the Johnny-Five servo object and its functions. Next, we built a project that taught us how to group servos with the servos object, and control them from the Pi's command line via REPL. Then, we built a project that showed the ability of the servo to convey information as well as create movement by building a light meter. Finally, we learned more about and played with continuous servos.

Questions

1. What are the differences between servos and motors?
2. What is the difference between regular and continuous servos?
3. Why do servos require an external power source?
4. When would you use servos over motors?
5. What are the benefits of the servos object?

8
The Animation Library

Servos are great tools for creating movement in our robotics projects but we need more control in order to create truly mobile walking robots. Each servo is different; for example, each servo moves at slightly different max speeds. If you want a bot to walk, you need timing control, and the ability to know when a servo has finished its movement. Enter the animation library; this powerful tool inside Johnny-Five allows you to fine-tune your servo movements to allow you more in-depth control.

The following topics will be covered in this chapter:

- Animating movement
- The terminology of the animation library
- The construction of the animation object
- Easing into your servo animations
- Learning more about queuing and playing animation segments
- Animation object events

Technical requirements

You'll need your two-servo setup from the previous chapter, and that's it for hardware for this chapter.

 The code for this chapter can be found at `https://github.com/PacktPublishing/Hands-On-Robotics-with-JavaScript/tree/master/Chapter08`.

Animating movement

The animation library makes many things possible with servos that are otherwise anywhere from difficult to downright implausible. Before we explore the *how* of the animation library, however, we should more thoroughly explain the *why*.

Why we need the animation library

Think about the movement of your leg as you take a step. You don't normally have to, but when you do, there's a lot going on in your joints! Your hip extends your leg out, and your knee extends your leg without usually locking it. And your back leg is doing things too; your hip is allowing the leg to move back, and your ankle is flexing. This is a massive oversimplification, but it's still really complicated!

Now imagine each of your joints as a servo, and you have to program taking a step. You cannot control the timing of each movement, because each servo will get to where you tell it to go as fast as you can. You also can't tell when a joint is done moving, so you have to hard-code timings and hope it holds up.

This exact kind of issue is what the animation library was made to alleviate. By giving you more control, you have more precision. But what do we mean by precision?

Moving servos with true precision

True precision, in the context of moving servos, means being able to control the timing, position, and speed of the servos being used. This level of precision is vital when building meticulous movements that require multiple movements happening in sync to avoid physical collisions. A great example of this is a hexapod robot: each joint needs to move in time with the other joints in the leg, and each leg needs to move at a precise time during a step in order to avoid colliding with each other or throwing the hexapod off balance.

Moving servos with true precision is a daunting task if you are hard-coding it; imagine setting 60 calls to `servo.to()` in order to create an animation that you *hope* takes one second. Or hard-coding each movement with `servo.to()`, timing it with the exact servo you've placed in the leg of your bot, and everything works...until the servo strips (it inevitably will), and you have to replace it and repeat the entire calibration process.

The animation library in Johnny-Five makes this process much simpler by allowing you to define your movements as segments of a larger design, that design being the animation itself. It does all of the math and works out all of the timings to ensure that your servos are where they need to be when they need to be there.

Implicit use of the animation library

Sometimes, you don't even need to create an animation object in order to create an animation for your servos. Before we really break into the animation object, let's write some code that uses the animation library implicitly.

Using servo.to() to implicitly create an animation

In a `project` folder for this chapter, create a file called `implicit-animations.js`. We'll set up this file to use the REPL to demonstrate animations created without explicitly creating an animation object. Start with the normal boilerplate: bring in `johnny-five` and `raspi-io`, set up your board and `board.ready()` handler as usual:

```
const Raspi = require('raspi-io')
const five = require('johnny-five')

const board = new five.Board({
  io: new Raspi()
})

board.on('ready', () => {
})
```

Then, inside the `board.on('ready')` handler, construct a `Servo` object:

```
let servo = new five.Servo({
  controller: "PCA9685",
  pin: 0
})
```

Next, while still inside the `board.on('ready')` handler, we're going to create the same motion three times—each time with a different set of specifications. Two of these will create an animation behind-the-scenes, and one is the default movement, so you can see the difference.

It can be a little hard to spot the differences in the servo movements if you're looking at a micro servo horn. I taped a Popsicle stick to my servo horns for this chapter, to make the differences easier to see. I also taped it to stand up on my desk so it wouldn't tip over, as shown in the following diagram:

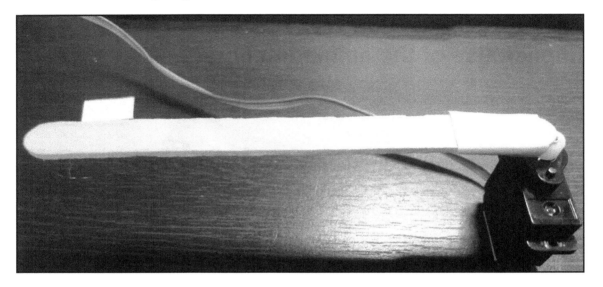

First movement function, still inside your `board.on('ready')` handler, will be called `normalFullSwing()`:

```
function normalFullSwing() {
 servo.to(0)
 servo.to(180)
}
```

This function will move the servo to 0 degrees as quickly as it can, then bring the servo to 180 degrees as fast as it can.

Let's add a parameter to `servo.to()` in our next function that will change how long the servo takes to get there. We'll set it to take in a time parameter that we'll pass through when we experiment in the REPL:

```
function timedFullSwing(time) {
  servo.to(0)
  servo.to(180, time)
}
```

This function will take the number of milliseconds passed to it to go from 0 to 180 degrees. It will still start by going to 0 degrees as fast as possible.

Finally, we'll write a function that takes in a time and a steps parameter that will move the servo from 0 to 180 in the time given, using the number of steps given. We'll call this function timedFullSwingWithSteps():

```
function timedFullSwingWithSteps(time, steps) {
  servo.to(0)
  servo.to(180, time, steps)
}
```

This function will, as the others, still go to 0 as fast as possible first.

Finally, we'll give ourselves access to these functions from the REPL using board.repl.inject():

```
board.repl.inject({
  normalFullSwing,
  timedFullSwing,
  timedFullSwingWithSteps
})
```

And we're ready to roll (or swing, as it were)!

Playing with implicit animations

Load the folder onto your Pi, navigate into it using your Pi SSH session, and run the following commands:

```
npm init -y
npm i --save johnny-five raspi-io
```

Now that we're all set up, we run the code using:

```
sudo node implicit-animations.js
```

Once you see Board Initialized, we can run our functions and see the differences. Here are just a few to try:

```
normalFullSwing() // hm...
timedFullSwing(1000) // wait a tick...
timedFullSwingWithSteps(1000, 10) // why didn't it do it?
```

You may be noticing something is up by now. At most, you might see a twitch or two, but it's certainly not working as intended!

That's because of a very important thing with timing, and that's that **if you don't wait for servo movements to finish, they'll just override each other, causing unstable results**.

That's part of why the animation library is so important! It has the ability for you to queue animations, meaning that the servo will let each movement finish before moving on to the next, preventing the need for you to program the waiting in yourself (especially icky considering the asynchronous nature of JavaScript).

Now, we're going to use a few more implicit animations, and some `setTimeout()` calls, to make these functions work properly.

Playing with implicit animations, take two

To fix your `normalFullSwing()` function, we'll set the `servo.to(0)` call to take 250 milliseconds, and call `servo.to(180)` after 255 milliseconds (just to be sure it's completely done getting to 0 first):

```
function normalFullSwing() {
 servo.to(0, 1000)
 setTimeout(() => { servo.to(180) }, 1010)
}
```

We'll do the same to the `timedFullSwing()` and `timedFullSwingWithSteps()` functions:

```
function timedFullSwing(time) {
  servo.to(0, 1000)
  setTimeout(() => { servo.to(180, time) }, 1010)
}

function timedFullSwingWithSteps(time, steps) {
  servo.to(0, 1000)
  setTimeout(() => { servo.to(180, time, steps) }, 1010)
}
```

Once you've made these changes, reload your `project` folder and run it:

```
sudo node implicit-animations.js
```

If you're using the book code directly instead of following along, `implicit-animations-fixed.js` contains the timeouts so you should run that file instead of running `implicit-animations.js` again.

Now that the code is running as intended, let's play around a little more with these implicit animations:

```
// Just a normal swing
normalFullSwing()
// Playing with the time parameter
timedFullSwing(1000)
timedFullSwing(750)
timedFullSwing(5000)
// Playing with the steps parameter
timedFullSwingWithSteps(3000, 2)
timedFullSwingWithSteps(1000, 10)
timedFullSwingWithSteps(1000, 100)
```

Make a mental note of what changing the timing and steps does to the servo's movements, it'll come in handy for the rest of the chapter.

Now that we understand some of the underlying effects of the animation library, and why it's so crucial when dealing with complex servo movements, let's dig into the animation library in detail. We'll start with the terminology, unless you have a strong background with animation (as in animated imagery), you have some vocabulary to learn!

The terminology of the animation library

The animation library was named the way it was quite intentionally; the vocabulary of the animation object very closely matches the vocabulary of animating images. Let's look at a few of the terms we'll be using heavily throughout this chapter.

- **Frame**: A frame of an animation is, in this context, the state of the servo at a given instance in time. As you can imagine, programming each and every frame of servo movement for a complex group of servos, such as a limb, would be a nightmare. Luckily, technology is on our side here, and we won't have to write each and every frame.
- **Keyframe**: A keyframe is a point in an animation that serves as an anchor unless you're drawing (or programming) every frame of an animation by hand; you establish a set of keyframes that establish the major points of movement for the animation. For example, in our full sweep we were doing earlier, a good set of keyframes would be something like this:
  - Start at any degree
  - Be at 0 degrees
  - End at 180 degrees

Simpler animations have fewer keyframes, but you always need at least two to create an animation. Note that keyframes themselves do not have any concept of time attached; they must be coupled with cue points to create an animation.

- **Cue point**: A cue point is a point in the context of the sequence between 0 and 1, and a set of cue points paired with an equally-sized set of keyframes and an overall duration creates a full animation. For instance, when we care the keyframes above with the set of cue points containing 0 seconds, 1 second, and 2 seconds, you get what starts to sound like an animation:
  - Start at any degree at 0%
  - Be at 0 degrees at 50%
  - End at 180 degrees at 100%

- **Duration**: Duration is the amount of time the animation sequence will run, and completes an animation when paired with keyframes and cue points. Take the above example with a duration of 2000 millseconds and you get:
  - Start at any degree at 0 ms
  - Be at 0 degrees at 1000 ms
  - End at 180 degrees at 2000 ms

- **Tweening**: Tweening is the idea of your software creating the necessary frames between your keyframes. You establish the keyframes, and tweening figures out what to do in between those frames. The time between each frame (exhibited by our `timedFullSwing()` function) and the number of steps (frames) between keyframes (exhibited by our `timedFullSwingWithSteps()` function) allow us to fine-tune the tweening process.

- **Easing**: Another part of the tweening process is easing. Without easing, all tweening is done linearly, with the same amount of movement in each tweened frame. This does not look smooth at all if you're building anything trying to walk. There are several forms of tweening; one of the most common form is ease-in or ease-out easing, which either starts slowly and ramps up to a fast ending, or starts fast and ramps down to a slow ending, respectively.

Now that we've discussed the terminology of an animation, we can start coding our first (explicit) animation object with Johnny-Five!

The construction of the animation object

To construct an animation object, we need to create the object itself, create a set of keyframes and a set of cue points, then enqueue those keyframes and cue points as an animation to run on our servos.

Creating the animation object

Create a new file in your `project` folder called `my-first-animation.js` and create the normal boilerplate: `require` in Johnny-Five and Raspi-IO, create your `Board` object, and create the `board.on('ready')` function:

```
const Raspi = require('raspi-io')
const five = require('johnny-five')

const board = new five.Board({
  io: new Raspi()
})

board.on('ready', () => {
})
```

Then, inside the `board.on('ready')` handler, construct our two `Servo` objects on pin 0 and pin 1 of our PWM hat:

```
let servoOne = new five.Servo({
  controller: "PCA9685",
  pin: 0
})

let servoTwo = new five.Servo({
  controller: "PCA9685",
  pin: 1
})
```

And create a `Servos` object containing our servos:

```
let servos = new five.Servos([servoOne, servoTwo])
```

Now that we have a group of servos, we can create an animation object:

```
let myFirstAnimation = new five.Animation(servos)
```

Now that we have our animation object, it's time to plan out our animation sequence, set keyframes and cue points, and queue them to animate.

Planning out the animation sequence

Let's plan a simple enough animation for our first go-round: let's allow the servos to start anywhere, then `servoOne` will move to 0 degrees while `servoTwo` will move to 180. Then, `servoOne` will sweep to 180 while `servoTwo` starts moving to 90 degrees, then both servos will end at 90 degrees. Let's have each of these positions happen two seconds apart. So our keyframes will look something like this:

1. Start with `servoOne` at any degree, start with `servoTwo` at any degree
2. Move `servoOne` to 0 degrees, move `servoTwo` to 180 degrees
3. `servoOne` stays at its last known position, `servoTwo` is moving towards 90 degrees
4. Move `servoOne` to 180 degrees, `servoTwo` is moving towards 90 degrees
5. Move `servoOne` to 90 degrees, move `servoTwo` to 90 degrees completed

Our cue points will be (in terms of 0-1): 0, .25, .5, .75, 1.

Now that we've planned out our sequence, we can start programming our keyframes.

Creating keyframes

We need to make a keyframe array for each servo in our servos group, for each cue point: two arrays of five keyframes each.

That sounds simple enough, but how do we tell the animation to let the servos start wherever they happen to be? And how do we tween `servoTwo` across two cue points in its move to 90 degrees? The answer lies in the way Johnny-Five parses null and false as servo positions in keyframes.

Using null and false as positions in keyframes

Null and false are used by Johnny-Five to allow us to create complex segments where we can tween between multiple cue points or use the last known position of a servo as a keyframe position.

The effect of null depends on where it is used, if it is used in the first keyframe, it uses the position of the servo as the animation begins as that keyframe's position. This is exactly what we need to start our animation sequence, as we want both servos to start at wherever they happen to be. If null is used in a keyframe that is not the first, then the keyframe will essentially be ignored at that cue point; if you have 30 in one keyframe, null in the next, and 120 in the third, the servo will move 90 degrees over the two cue points. We will use this to allow servoTwo to move from 180 to 90 over two cue points.

When you use false as a keyframe position, it will use the position set in the last keyframe. We will use this on servoOne when the keyframe calls for the servo to remain in its last known position, instead of hard-coding a second 180-degree position.

Now that we know how null and false affect our positioning in keyframes, let's program our keyframes for our planned animation sequence.

Programming our keyframes

So based on the information we've been given, the values we need for each keyframe are:

- servoOne null, servoTwo null (start wherever the servos happen to be)
- servoOne 0, servoTwo 180
- servoOne 180, servoTwo null (servoTwo starts moving towards 90 degrees)
- servoOne false, servoTwo null (servoOne stays put, servoTwo still moving to 90 degrees)
- servoOne 90, servoTwo 90

Each position needs to be an object with a property degrees for each keyframe. Let's translate that into JavaScript, right under the construction of our animation object:

```
let keyframes = [
  [null, {degrees: 0}, {degrees: 180}, false, {degrees:90}], // servoOne
  [null, {degrees: 180}, null, null, {degrees: 90}] // servoTwo
]
```

Now that we have our keyframes programmed, let's get started on our cue points.

Setting cue points and duration

Cue points, no matter how many servos you have, will always be a one-dimensional array of times to match each keyframe in the array of keyframes you pass in.

Note that while the cue points in this animation are evenly spaced, that is absolutely optional your cue points can vary wildly in distance from each other without breaking anything.

Underneath our keyframes object, let's set up our cue points array:

```
let cuePoints = [0, .25, .5, .75, 1]
```

We want our animation to take 8 seconds overall, so add:

```
let duration = 8000
```

We have all the data we need, let's make an animation!

Putting it all together to make an animation

In order to run our animation sequence, we have to place it in the queue using the `Animation.enqueue()` function. We'll need to pass in the duration, keyframes, and cue points together. In your `my-first-animation.js`, after the duration, add:

```
myFirstAnimation.enqueue({
  duration: duration,
  keyFrames: keyframes,
  cuePoints: cuePoints
})
```

The object containing the duration, `keyFrames`, and `cuePoints` properties is known within the animation library as a `Segment` object.

The animation segment will immediately begin to play upon queuing, so we're ready to load our project in and see some animated servos!

Watching your animation at work

Load your `project` folder onto the Pi, navigate into the folder in your Pi SSH session, and run:

```
sudo node my-first-animation.js
```

You should see the animation play out with the two servos as we described it.

This is really powerful, but when you think of a walking hexapod, these linear movements wouldn't make for a realistic or pretty gait. Let's add some easing into our animation sequence in order to create some more organic-looking movement.

Easing into your servo animations

Unless you want any of your future walking bots to be very firmly in the uncanny valley, you'll need to use easing to create a more fluid, natural motion with your animation segments.

How easing fits into an animation segment

Easing functions are added into the `keyframes` of a servo; so not only are we saying what position we want the servo to be, but how it gets there. For example, these `keyframes`:

```
let keyframes = [
  null,
  {degrees: 180, easing: 'inoutcirc'}
]
```

Will take a servo starting at any position and move it to `180`, starting out slow, speeding up in the middle, and slowing down again towards the end.

There are many different options for easing, and they are documented in the ease-component (`https://www.npmjs.com/package/ease-component`) npm module included as a dependency to Johnny-Five. We'll be using `incirc`, `outcirc`, and `inoutcirc` to start.

Adding easing to our first animation

Copy the contents of `my-first-animation.js` into a new file called `easing-animations.js`. Next, we'll modify the `keyframes` array to include some easing:

```
let keyframes = [
  [null, {degrees: 0}, {degrees: 180, easing: 'inOutCirc'}, false,
{degrees:90, easing:'outCirc'}], // servoOne
  [null, {degrees: 180}, null, null, {degrees: 90, easing:'inCirc'}] //
servoTwo
]
```

Let's also increase the duration of the animation segment so we can really see the difference easing makes:

```
let duration = 16000
```

Then, load it onto the Pi, navigate to the folder in your Pi SSH session, and run the following command:

```
sudo node easing-animations.js
```

Really watch how `inCirc`, `outCirc`, and `inOutCirc` affect your animation.

Easing an entire animation segment

In order to easily set all keyframes in an animation segment to have the same easing, you can pass an `easing` property when you enqueue your segment. For example:

```
myFirstAnimation.enqueue({
  keyFrames: keyframes,
  duration: duration,
  cuePoints: cuePoints,
  easing: 'inOutCirc'
})
```

The preceding code will override the keyframes and all of them will use `inOutCirc` easing. Now that we've fully explored easing our animation segments, let's take a look at the animation queue and how we can affect our segments when we queue them and when they're playing.

Learning more about queuing and playing animation segments

When we queue an animation segment, we pass it a duration, cue points, and keyframes. But we can also pass in other options that affect the playback of the animation segment. We can also call methods on the animation object that affect animation segments currently playing and in the queue.

Before we start messing with these, copy the content of `easing-animations.js` into a new file called `playing-with-the-queue.js`. Remove the call to `myFirstAnimation.enqueue()` at the end; we want a little control when we get into the REPL this time around.

Looping animation segments

First, let's add a function that will enqueue our animation normally:

```
function playMyAnimation() {
  myFirstAnimation.enqueue({
    keyFrames: keyframes,
    duration: duration,
    cuePoints: cuePoints
  })
}
```

Sometimes you want the animation segment you are enqueuing to run on a loop. Let's create a function in our `board.on('ready')` handler that will enqueue our animation segment on a loop:

```
function loopMyAnimation() {
  myFirstAnimation.enqueue({
    keyFrames: keyframes,
    duration: duration,
    cuePoints: cuePoints,
    loop: true
  })
}
```

You can also add the `loopBackTo` property, and set it to the index of a cue point; the animation will start its loop from the designated cue point.

What if we want the animation to play forward, then back to the start, and repeat? Let's write a function that will set the `metronomic` property to do just that:

```
function metronomeMyAnimation() {
  myFirstAnimation.enqueue({
    keyFrames: keyframes,
    duration: duration,
    cuePoints: cuePoints,
    metronomic: true
  })
}
```

Now that we know how to loop and metronome our animation segments, let's explore changing the speed of animation segments using the animation object.

Changing the speed of animation segments

You can call `Animation.speed()` with a numeric multiplier to change the speed of the current animation segment. For instance, calling `Animation.speed(.5)` will halve the speed, and `Animation.speed(2)` will double it.

Let's write some functions to half, double, and normalize our animation segment speed:

```
function halfSpeed() {
 myFirstAnimation.speed(.5)
}

function doubleSpeed() {
 myFirstAnimation.speed(2)
}

function regularSpeed() {
 myFirstAnimation.speed(1)
}
```

Add these to the loop and metronome functions.

Now that we know how to adjust the speed of animation functions, as well as how to loop them, let's talk about pausing, playing, and stopping animations.

Playing, pausing, and stopping animation segments

If left alone, the animation segments will be played until there is nothing left in the queue to play (meaning if there is a looped or metronome segment, it will stay on that segment).

But you can move to the next animation:

```
Animation.next()
```

Or you can pause the current segment:

```
Animation.pause()
```

Start it up again:

```
Animation.play()
```

Or stop the current segment and clear out the entire queue:

```
Animation.stop()
```

Let's use these, along with the REPL, to play with our animation and our new-found powers to manipulate it.

Tying it all together in the REPL

Add the following to the end of the `board.on('ready')` handler in `playing-with-the-queue.js`:

```
board.repl.inject({
    myFirstAnimation,
    playMyAnimation,
    loopMyAnimation,
    metronomeMyAnimation,
    halfSpeed,
    doubleSpeed,
    normalSpeed
})
```

Then, load your `project` folder onto your Pi, navigate to the `project` folder in your Pi SSH session, and run the following command:

```
sudo node playing-with-the-queue.js
```

Once you see `Board Initialized`, try a few commands to experiment with how your animation plays:

```
loopMyAnimation()
myFirstAnimation.pause()
myFirstAnimation.play()
halfSpeed()
myFirstAnimation.stop()
metronomeMyAnimation()
doubleSpeed()
playMyAnimation()
myFirstAnimation.next() Summary
```

Summary

In this chapter, we dived into the animation library with servos. We learned the key terminology for the animation library, how to construct an animation segment, how to queue it, and how to manipulate playback, both when queuing the segments or by calling methods of the animation object.

Questions

1. Why are animations necessary for complex movements with multiple servos?
2. What is a keyframe?
3. What is a cue point?
4. Name the three pieces of an animation segment.
5. What does easing do to our animation keyframes and segments?
6. What method of the animation object stops the current segment and clears the animation queue?
7. What does calling `Animation.speed(.25)` do to the current animation?

Getting the Information You Need 9

We've let our Pi discover its immediate surroundings, and let it show data through various means. We've even given them the ability to move! But there's a cosmos of data to be collected, and sometimes the data you want can't be collected locally. That's where the internet, and initiatives to make more and more data freely available, come into play. In this chapter, we'll look into connecting your Pi to the internet and obtaining weather information in order to create a weather dashboard.

The following topics will be covered in this chapter:

- Why connect your NodeBots to the internet?
- Getting weather data on our Pi with OpenWeatherMap
- Building a weather dashboard with an LCD
- Scraping websites for data with your Pi

Technical requirements

For this project, you'll want your Pi and an LCD character display with an I^2C interface. You can purchase and solder together an LCD (https://www.adafruit.com/product/198) and backpack (https://www.adafruit.com/product/292) from Adafruit, or a pre-built module via SainSmart (https://www.amazon.in/SainSmart-Serial-Module-Shield-Arduino/dp/B00AE0FRDQ/).

You'll also want to make sure your Pi can access the outside world with internet access, as we set it up to in Chapter 1, *Setting Up Your Development Environment*.

The code for this chapter is available at: `https://github.com/PacktPublishing/Hands-On-Robotics-with-JavaScript/tree/master/Chapter09`.

Why connect your NodeBots to the internet?

While sensors can provide local data, sometimes you want to display data from far away or data from sensors attached to other devices. This is where we can really leverage Node.js and npm packages in our favor for our Raspberry Pi projects.

Using the power of npm modules

Back in `Chapter 2`, *Creating Your First Johnny-Five Project*, we used the `color` npm module to manage colors for us. We've used the `barcli` module to get our sensor data into bar graphs. Now it's time to use the request npm module to retrieve data from websites for us! This allows us to simplify development over microcontrollers that use C by not having to create HTTP requests by hand each time, and being able to use asynchronous calls.

For those unfamiliar with the request module, we'll use it to make HTTP GET requests like so:

```
const request = require('request')

request.get(url, (err, response, body) => {
  console.log(body)
})
```

We give the `request.get()` call a URL and a callback that receives an error (that is, hopefully, null), a response object, and a body which is conveniently extracted for us from the full request object (which can be huge and complex).

Using the data you collect

You can use data you collect from the internet for many different projects:

- I have a string of lights in my lab that are controllable by Twitch live chat
- You can compare information in a local project to data from far away
- You can just use random data! Markov chains and other semi-random data can make for fun projects

Just a few things you'll want to know that will be going into your data collection project:

- Is this a REST API? Will I get JSON data back or will it need to be parsed?
- Is this scraping an HTTP website? How will I parse out the HTML data I'm looking for? (Caution: this gets tricky and can be brittle if the website you're scraping changes often.)
- Do I need an API key or JSON Web Token (JWT) for authentication purposes?

Some things to keep in mind

Here are some things to keep in mind when doing internet data collection on your Pi projects:

- Wi-Fi uses a lot of power, so if you're running your project on a battery, you'll need to keep power consumption in mind.
- Use your robotics powers for good, don't build projects that do harm, collect information they shouldn't, or have other dubious purposes!
- Parsing out huge JSON or HTML responses can take a while on the Pi, so take a look at what you're getting if your project is running a bit slowly.

Getting weather data on our Pi with OpenWeatherMap

We're going to build a weather bot for this one, and while we could use a temperature sensor, that'd only tell us what it's like indoors, and usually we'd like to see what the weather's like outside before we head out the door. So we're going to use the OpenWeatherMap API to get data and display it on a character LCD; but let's walk before we can run by starting with getting the data from the API to the Pi.

Getting an OpenWeatherMap API key

First, you'll need to sign up for an account at `https://openweathermap.org/`, and generate an API key. Then, click your username in the upper-left corner and select **API Keys** from the tabs that appear near the top of the page:

Generate an API key on this page and keep the tab open; we'll use it in the next section to get the info we need.

Next, open the API link in a separate tab; you'll see the main API function calls on this page. The one we're looking for is right at the top – current weather data.

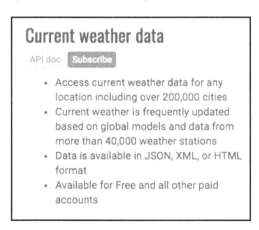

Click the **API doc** button and we'll figure out what URL we'll need to make a request to. At the time of writing, the URL is as follows:

```
http://api.openweathermap.org/data/2.5/weather?q=[city]&appid=[your API
key]
```

If you go to that URL in your browser with your city and API key filled in, you should see something like this:

```
{"coord":{"lon":-97.74,"lat":30.27},"weather":[{"id":800,"main":"Clear","de
scription":"clear
sky","icon":"01n"}],"base":"stations","main":{"temp":305.59,"pressure":1016
,"humidity":46,"temp_min":304.15,"temp_max":307.15},"visibility":16093,"win
d":{"speed":3.6,"deg":170,"gust":8.2},"clouds":{"all":1},"dt":1534470960,"s
ys":{"type":1,"id":2558,"message":0.0037,"country":"US","sunrise":153450715
6,"sunset":1534554612},"id":4671654,"name":"Austin","cod":200}
```

> If those temperatures look a little high to you (even for Austin, TX), that's because they're in degrees Kelvin. We'll pass the units parameter in our URL in our project set to metric for degrees Celsius or 'imperial' for degrees Farenheit.

Now we're ready to code in the npm request module and get our data into our Pi.

Bringing in request

Let's write a basic program without Johnny-Five to collect our data on the Pi before adding in our LCD. In a file on your Pi or ready to be moved to your Pi called weather-test.js:

```
const request = require('request')

setInterval(() => {
  request({
    url: 'http://api.openweathermap.org/data/2.5/weather',
    qs: {
      q: [your city],
      appid: [your API key],
      units: ['metric' or 'imperial']
    },
    json: true // returns the parsed json body for us
  }, (err, resp, body) => {
    console.log(body)
  })
}, 60000)
```

Parsing the response

The JSON object that is printed out by the console looks something like this (formatted for easier reading):

```
{
  "coord":{"lon":-97.74,"lat":30.27},
  "weather":[
    {"id":800,"main":"Clear","description":"clear sky",
      "icon":"01n"}
  ],
  "base":"stations",
  "main":{
    "temp":305.59,
    "pressure":1016,
    "humidity":46,
    "temp_min":304.15,
    "temp_max":307.15
  },
  "visibility":16093,
  "wind":{
    "speed":3.6,
    "deg":170,
    "gust":8.2
  },
  "clouds":{"all":1},
  "dt":1534470960,
  "sys":{
    "type":1,
    "id":2558,
    "message":0.0037,
    "country":"US",
    "sunrise":1534507156,
    "sunset":1534554612
  },
  "id":4671654,
  "name":"Austin",
  "cod":200
}
```

That's a lot of weather data! Luckily, because the request npm module was passed the `json: true` property in the options, it assumes that anything passed back is JSON and parses it for you, so you can access data properties right away:

```
let longitude = body.coords.lon // -97.74
let conditions = weather.condition // 'clear sky'
let currentTemp = main.temp // 305.59 degrees Kelvin
```

Building a weather dashboard with an LCD

Now that we have our weather data, it's time to wire our character LCD to our Pi and use it to show the weather data. We'll explore the Johnny-Five LCD object, wire it to the Pi, and code it all together with Johnny-Five and the npm request module.

Adding an LCD to the Pi

Refer to the following diagram for the connection:

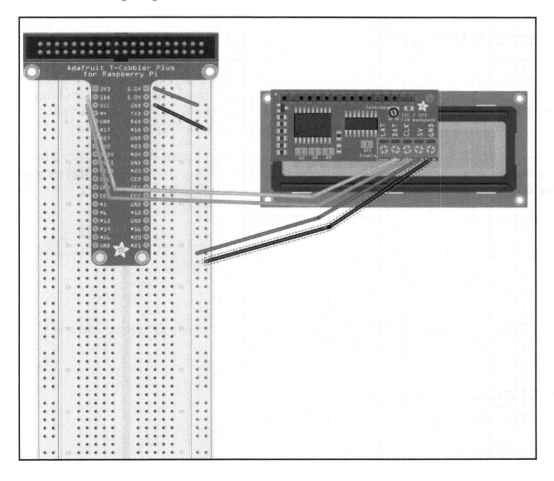

Keep in mind that backpack with the I²C interface is on the back of the LCD; I moved it forward in the diagram to help you see the connections to the Pi.

The LCD object

Let's take a look at the LCD object in the Johnny-Five documentation in order to figure out how to construct and use our LCD in our weather dashboard code.

Constructing our LCD

Usually, an LCD without the I2C can take up to eight pins! That's a lot, and I like as few wires as possible in my robotics projects (easier to debug later). With our backpack, we only need the two power pins and two I2C pins. But that also means we'll need to find our controller—if you are using the Adafruit backpack, then our controller is the PCF8574; if you're bought another backpack, make sure it uses one of the PCF8574x chips!

We'll also need the size of the LCD in rows and columns of characters—most are 2 rows by 16 columns, but you may have gone bigger with the 4 row by 20 columns character model. In either case, use whichever works for the LCD you attached to the backpack.

With all that in mind, our constructor should look like this:

```
let LCD = new five.LCD({
  controller: "PCF8574",
  rows: 2,
  cols: 14
});
```

Now that we've constructed our LCD, let's see what we need to get set up and get characters on the screen!

Setting up the LCD

First thing we need to do is turn the backlight on:

```
LCD.on()
```

Then, we'll want to make the blinking cursor disappear:

```
LCD.noBlink()
```

Now, we're ready to learn about moving the cursor, printing statements, and clearing the LCD.

Printing to and clearing the LCD

Before we print, we want to make sure that the LCD is cleared out:

```
LCD.clear()
```

And that the cursor is in the home (row 0, col 0) position:

```
LCD.home()
```

Next, we can print to the LCD:

```
LCD.print("Hello, World!")
```

Note, you can also chain LCD functions together, as Johnny-Five returns the LCD object from every object function:

```
LCD.clear(),home(),print('Hello World!')
```

Now, we have everything we need to start making our dashboard!

Coding it all together

We need to take what we've learned in this chapter and put it together. Start by creating a file called `weather-dashboard.js` in your `project` folder, and setting up your Johnny-Five, Raspi-IO, and request libraries, constructing your `Board` object, and creating your `board.on('ready')` handler:

```
const Raspi = require('raspi-io')
const five = require('johnny-five')
const request = require('request')

const board = new five.Board({
 io: new Raspi()
})

board.on('ready', () => {
})
```

Then, inside the `board.on('ready')` handler, construct and set up our LCD:

```
let LCD = new Five.LCD({
  controller: 'PCF8574'
})

LCD.noBlink()
LCD.on()
```

Then, we'll create a function that gets the weather data, and set it on an interval of one minute:

```
function getWeather() {
  request({
    url: 'http://api.openweathermap.org/data/2.5/weather',
    qs: {
      q: [your city],
      appid: [your API key],
      units: ['metric' or 'imperial'],
      json: true
    }
  }, (err, resp, body) => {

  })
}

setInterval(getWeather, 60000)
```

In the request callback, we'll clear, and write to, the LCD:

```
LCD.clear()
LCD.home().print('Temp: ' + body.main.temp + ' Deg [F or C]')
LCD.setCursor(0, 1).print(body.weather.description)
```

Finally, call the `getWeather()` function at the start to prevent the project from taking a full minute before showing anything:

```
getWeather()
```

Once you have the full code together, load the `project` folder onto our Pi the following, navigate to the folder in your Pi SSH session, and run the following commands:

```
npm i
sudo node weather-dashboard.js
```

You should have the temperature and conditions for the city you put in appear on the LCD, and they should refresh every minute.

Now that we've seen a project where the Pi pulls from a nice neat JSON REST API, let's take a crack at getting data from a bit more difficult source: HTML scraping.

Project – scraping data from websites with your Pi

HTML scraping is the process of making a request to a webpage in order to obtain the HTML itself, so data can be parsed out of it. We're going to build a bot that shows whether `johnny-five.io` is up or not by scraping `https://downforeveryoneorjustme.com/`, a site that tells you if a site is down from multiple sources.

You don't need to change the wiring setup from the weather dashboard for this project, our current hardware is all we need.

Scraping downforeveryoneorjustme.com for johnny-five.io

First, go to `https://downforeveryoneorjustme.com/` and enter `johnny-five.io` in the URL input, and hit *Enter*. You should end up at `https://downforeveryoneorjustme.com/johnny-five.io`, where hopefully you'll see a rather simple page that looks like this:

Now to prepare for our web scraping code, we need to know what HTML element we're looking for as well as the URL. Right-click the **It's just you.** and select **Inspect** (or whichever variation it is on your browser. In Chrome, you'll see something like this:

```
▼<div id="domain-main-content">
    <h1 id="seo-h1">Is johnny-five.io down?</h1>
  ▼<p> == $0
      "
      It's just you.   "
      <a href="https://johnny-five.io" class="domain">johnny-five.io
      </a>
      " is up.
      "
    </p>
  ▶<ul>...</ul>
    <p></p>
  ▶<div id="showcontent">...</div>
  </div>
  <div id="domain-side-actions">
```

This is one of the perils of web scraping: there's not always a lot to go on to find your element. The closest we get is the first paragraph (p tag) of the div with ID domain-main-content. We'll want to see if it contains the string It's just you. in order to determine if johnny-five.io is up.

Now that we have our URL and intended element and parsing criteria, let's start coding by getting the HTML into our Johnny-Five project.

Making the HTTP request

Create a new file in your project folder called scraper-j5-alert.js. Start with the normal libraries, Board construction, and board.on('ready') handler. Don't forget to include the request npm module:

```
const Raspi = require('raspi-io')
const five = require('johnny-five')
const request = require('request')

const board = new five.Board({
  io: new Raspi()
})
```

```
board.on('ready', () => {
})
```

Then, inside your `board.on('ready')` handler, construct and set up your LCD object:

```
let LCD = new Five.LCD({
 controller: 'PCF8574'
})

LCD.noBlink()
LCD.on()
```

Then, we're going to create a function to get the HTML from `https://downforeveryoneorjustme.com/johnny-five.io` and place it on a five-minute interval. Lastly, we call it so we don't have to wait five minutes for the first result:

```
function isJohnnyFiveDown() {
    request('https://downforeveryoneorjustme.com/johnny-five.io',
      (err, resp, body) {
        console.log(body)
      })
}

setInterval(isJohnnyFiveDown, 300000)
isJohnnyFiveDown()
```

Load the `project` folder onto your Pi, navigate to the folder in you Pi SSH session, and run:

sudo node scraper-j5-alert.js

You should, maybe after a few seconds, see something like this in the console (I just screenshotted a small part):

```
<section id="tabs-1">
   <div id="domain-main-content">
     <h1 id="seo-h1">Is johnny-five.io down?</h1>

     <p>
It's just you.    <a href='https://johnny-five.io' class="domain">johnny-five.io</a></span>
 is up.
</p>
```

But how do we get the info out of that giant string? Regular Expressions? Please, no, not those. Luckily, as Stilwell's Law (see `Chapter 2`, *Creating Your First Johnny-Five Project*) states, if you can think of functionality, there exists a package on `npm` for it. In this case, we have the cheerio module that allows us to parse and query the HTML string with a JQuery-style API.

Using Cheerio to get the element we want

In your `project` folder, run the following command:

```
npm i --save cheerio
```

The basics on cheerio are you parse text by calling:

```
const cheerio = require('cheerio')
const $ = cheerio.load(htmlText)
```

Then, query using the `$` variable like you would with JQuery (see *Further reading* if you've never used JQuery for a link to a great primer on selecting elements):

```
let divWithIDHello = $('#hello')
let helloDivText = divWithIDHello.text()
```

Parsing the HTML and showing the result

This is all we'll need to scrape the HTML and get the status. Inside `scraper-alert-j5.js`, we're going to add cheerio's `require()` call to the top of the file with the other requires:

```
const cheerio = require('cheerio')
```

Then, we're going to modify the callback that fires when request is done fetching the HTML. We're going to add the cheerio call to load the text and look for the first p child of the div with ID `domain-main-content`, and pull out its text. Then, we'll see if that text contains `It's just you.` and write to the LCD:

```
function isJohnnyFiveDown() {
  request('https://downforeveryoneorjustme.com/johnny-five.io',
    (err, resp, body) => {
      let $ = cheerio.load(body)
      let statusText = $('#domain-main-content p')[0].text()
      LCD.clear()
      LCD.home()
```

```
        LCD.print('johnny-five.io')
        LCD.cursor(0, 1)
        // make sure to use " to surround the string!
        if(statusText.contains("It's just you.")){
          LCD.print('is up!')
        } else {
          LCD.print('is down (possibly)!')
        }
    })
  }
```

We're ready to load it and run it! Load your project onto your Pi, navigate to the folder in your Pi SSH session, and run the following commands:

```
npm i
sudo node scraper-alert-j5.js
```

And you should see whether Johnny-Five is up or not on your LCD!

You may have noticed I put `possibly` in the down condition. This is because, as I mentioned before, HTML scraping is very brittle. If they change **It's just you.** to **It is just you.**, our code will break! So I like to remind the LCD viewer that it may not necessarily be down. Again, this is an example of why, if you can find it, it's better to get data from an API. But sometimes there's no real choice.

Summary

In this chapter, we built a weather dashboard using an I²C LCD screen knowledge of npm modules and using REST APIs, and leveraged the power of Node.js and the Pi together. You can go forward to build so many new projects with these skills; if you can get the information from the internet, you can use it in your Johnny-Five and Pi projects.

Questions

1. Why is the Pi well suited for projects that require remote data?
2. What considerations need to be made when making regular web requests from the Pi?

2. Why can we chain the LCD object calls, such as `LCD.clear().home()`?
3. Why do we use an I2C backpack with our LCD?
4. Would we need more components to use the LCD without the backpack?
5. Does `LCD.on()` turn on the entire LCD? If not, what does it do?

Further reading

- **The request npm module page**: `https://www.npmjs.com/package/request`
- **The full OpenWetherMap API**: `https://openweathermap.org/api`
- **The Johnny-Five LCD documentation**: `http://johnny-five.io/api/lcd/`
- **The cheerio npm module page**: `https://www.npmjs.com/package/cheerio`
- **JQuery 'Selecting Elements' tutorial**: `https://learn.jquery.com/using-jquery-core/selecting-elements/`

10
Using MQTT to Talk to Things on the Internet

IoT devices can communicate in many ways, and some ways have become standards. We're going to explore a few of the ways IoT devices communicate, then dive into depth with one standard, MQTT. We'll then build a small project that allows us to see and send MQTT events with AdafruitIO, a service that provides MQTT brokers online.

The following topics will be covered in this chapter:

- IoT device communications
- MQTT – an IoT PubSub protocol
- Setting up MQTT on the Pi with AdafruitIO
- Project – adding an LCD and button to see and show MQTT events
- Project – social media notification bot with IFTTT

Technical requirements

For this project, you'll need your LCD hooked up to your Pi (see Chapter 9, *Getting the Information You Need*), a pushbutton, and a 10K ohm resistor for hardware.

For other tools, you'll want to create a free-plan account at https://adafruit.io. You will also want an account with IFTTT (also free) to do the final project.

 The code for this chapter is at https://github.com/PacktPublishing/ Hands-On-Robotics-with-JavaScript/tree/master/Chapter10.

IoT device communications

As we saw in Chapter 9, *Getting the Information You Need*, our Pi can ask for information from the internet using HTTP requests. But what if we want regular data sent to the Pi in real-time? What if we want a swarm of devices chatting with each other, data sent back and forth as necessary? Let's take a look at a few ways this can be accomplished with web technologies.

Long polling

Long polling involves asking for information via HTTP requests at certain intervals. If this sounds familiar, it's because that's precisely what we did in our weather dashboard project in Chapter 9, *Getting the information you need*; we poll the OpenWeather API every 60 seconds. This approach is best when there aren't other options; some REST APIs do not have a way to hold a connection open or establish two-way communication, and long polling is the way to go in these situations.

But there are newer ways of establishing two-way connections that can be left open, including the Websocket.

Websockets

Websockets are a powerful tool that allows us to establish a two-way data connection that stays open until you close it (barring error or loss of connection). You can send messages back and forth in real time, without having to set up a whole new connection each time. It also allows the server to communicate back with your Pi without it having to ask, which is great for real-time data.

While websockets are powerful, there are ways to fine-tune this connection for IoT projects. It can also be difficult to maintain an open socket and keep the data you are sending or receiving organized. With this in mind, we're going to talk about MQTT.

MQTT - an IoT PubSub protocol

The **Message Queuing Telemetry Transport** (**MQTT**) protocol (often pronounced **mosquitto**) is a protocol designed for low-bandwidth, high-latency environments, which makes it a great fit for IoT projects, especially ones running on limited hardware. It is not solely used for communication between machines: some projects use it to send data for storage purposes.

Let's take a look at how MQTT works and how it implements a PubSub setup for our projects.

The basics of MQTT

Let's go over a few terms, then link them together to define MQTT as a concept:

- **MQTT message**: An MQTT message consists of a topic and a message. The topic is what clients subscribe to, and they usually read the messages for data.
- **MQTT client**: An MQTT client connects to an MQTT broker and can subscribe to and publish on topics once connected to the broker.
- **MQTT broker**: An MQTT broker handles client connection and passes messages along to all clients subscribed to a topic when a client publishes a message on that topic. It can also publish messages to topics, which go to all clients subscribed.

In our project, we will set up our Pi as an MQTT client, connect it to a broker at AdafruitIO, and publish messages that the broker will send to the client (our Pi) and takes in messages it publishes.

Setting up MQTT on the Pi with AdafruitIO

In order to set up MQTT, we'll need a broker. While the Pi can itself serve as a broker (see the *Further reading* section), we don't need broker functionality on the Pi. We can use AdafruitIO to create a broker that we can subscribe to topics with on our Pi.

Creating an account and a feed

First, we'll go to `https://io.adafruit.com/` and create a free account. Then, you'll be taken to a dashboard:

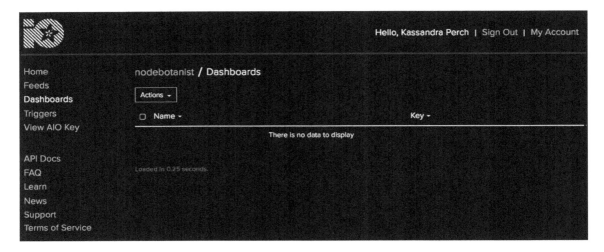

We'll need to set up a feed and get our AIO key in order to get started on the Pi. To set up a feed, select **Feeds** in the left-hand menu. Then, click the **Actions** button in the upper left corner:

Next, select **Create New Feed** from the dropdown. Name your feed (I named mine `hands-on-robotics-with-js` to make it easy to remember what I created it for). Then, you'll be taken to your new feed's page:

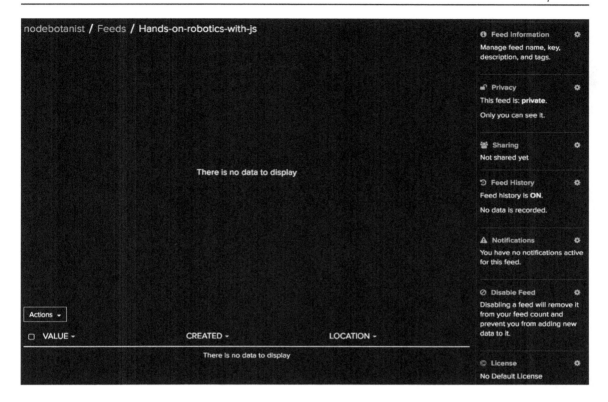

Now that we've created our feed, we can get all the information we need to get the Pi hooked up to it. First, let's get the Feed key to give to the MQTT module. Click **Feed Information** on the right-hand side and copy the value in **MQTT by ID** and place it somewhere it's easy to get to. Then, close the Feed Information window.

Next, click **View AIO key** in the left-hand menu. Copy the value in **Active Key** and place it somewhere it's easy to get to.

Now, to clear up some terminology before we proceed: the feed **MQTT by ID** name you took note of is going to be the topic of any messages our client (the Pi) sends to our broker (AdafruitIO), and vice versa.

We now have everything we need to connect our Pi to the AdafruitIO broker.

Subscribing to the feed using the mqtt npm module

Create a `project` folder for this project to be transferred to the Pi. Inside the folder, run the following commands:

```
> npm init -y
> npm i --save mqtt dotenv
```

This will install the `mqtt` npm module, which will simplify our MQTT connection, and the `dotenv` module, which will allow you to use environmental variables in a separate file (great for making sure you don't commit your AIO key to GitHub!)

The mqtt module

The `mqtt` module allows us to construct an MQTT client object:

```
const client = mqtt.connect(url, {options})
```

We will pass this method the AdafruitIO URL and our username and AIO key via the `options` object.

There are several event handlers available on the MQTT client object (see *Further reading*), but for this project we'll be using the following:

```
client.on('connect', () => {})
client.on('message', (topic, message) => {})
```

Finally, we'll want to be able to publish messages from our client to the broker:

```
client.publish(topic, message)
```

The dotenv module

The `dotenv` module makes configuration of environment variables that you don't want to, say, commit to GitHub, easy. You load it into your Node.js application:

```
const dotenv = require('dotenv').config()
```

This loads a `.env` file in the same directory that the Node.js file is in, which is in the following format:

```
KEY=value
```

And it is then accessible in your application via the `process.env` global variable:

```
let key = process.env.KEY // 'value'
```

Now that we know more about the libraries and services we're using, let's get a test connection program set up!

Testing our connection

To test our connection, we'll get our Pi connected to the AdafruitIO broker, subscribe to our new feed, and publish a message. We'll know it works when we go to the feed dashboard on the AdafruitIO site and see that our message has been received.

To do this, we need to configure the `mqtt` client, set up a `connect` handler for the client, and use that handler to subscribe and publish our message. In a file called `mqtt-test.js` in the folder you set up earlier, write the following code:

```javascript
const dotenv = require('dotenv').config()
const mqtt = require('mqtt')

const client = mqtt.connect(
  process.env.ADAFRUIT_IO_URL,
  {
    username: process.env.ADAFRUIT_IO_USERNAME,
    password: process.env.ADAFRUIT_IO_KEY,
    port: process.env.ADAFRUIT_IO_PORT
  }
)

client.on('connect', () => {
  console.log('Connected to AdafruitIO')
  client.subscribe(process.env.ADAFRUIT_IO_FEED, () => {
    client.publish(process.env.ADAFRUIT_IO_FEED, 'Hello from the Pi!')
  })
})
```

Then, in the same folder, create a file called .env (make sure it starts with the dot!), and place the following in it:

```
ADAFRUIT_IO_URL=
ADAFRUIT_IO_USERNAME=[the username you signed up for AdafruitIO with]
ADAFRUIT_IO_KEY=[your AIO Key from earlier]
ADAFRUIT_IO_PORT=
ADAFRUIT_IO_FEED=[The feed info from earlier]
```

Then, transfer the folder over to the Pi. In the Pi session, navigate to the folder and run the following command:

```
npm i
```

The preceding command will make sure that all modules are installed correctly on the Pi. Then, use the following:

```
node mqtt-test.js
```

Now go to your AdafruitIO feed dashboard. You should see a message there:

Now that we know we can connect the Pi to AdafruitIO, let's add an LCD to see incoming messages, and a button to generate outgoing ones!

Project – adding an LCD and button to see and send MQTT events

We can use the AdafruitIO dashboard to post messages to our MQTT feed, and so we'll use an LCD to show what we've sent. We'll also wire up a button that will send an MQTT message when pushed.

Wiring it all up

First, we'll wire our LCD to the I2C pins, and our button to GPIO #5, also known as P1-29:

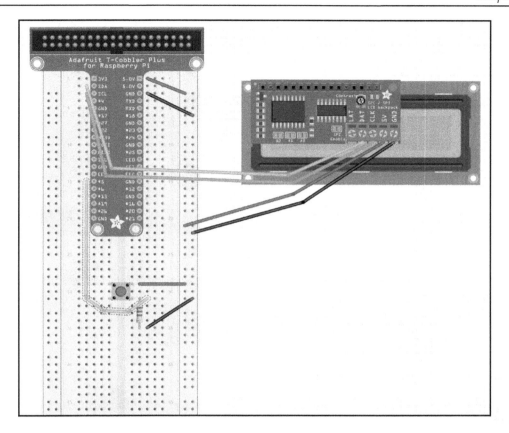

Coding it all together

In a file in the same folder, create `mqtt-button-lcd.js`. Put in the usual Johnny-Five and Raspi-IO constructors, and in the board-ready handler:

Then, add the client constructor for AdafruitIO's MQTT connection from `mqtt-test.js`. We'll also set up our LCD and button objects here:

```
let LCD = new five.LCD({
  controller: "PCF8574",
  rows: 2,
  cols: 16
})
let button = new five.Button('P1-29')
```

After that, we're ready to code the sending of messages on the press of the button, and the printing of messages received on the LCD:

```
client.on('connect', () => {
  console.log('Connected to AdafruitIO')
  client.subscribe(process.env.ADAFRUIT_IO_FEED, () => {
    client.publish(process.env.ADAFRUIT_IO_FEED, 'Hello from the Pi!')

    button.on('press', () => {
      client.publish(process.env.ADAFRUIT_IO_FEED, 'Button pressed!')
    })

    client.on('message', (topic, message) => {
      LCD.clear().home().print(topic).setCursor(1,0).print(message)
    })
  })
})
```

Now, move the folder over to the Pi, go into your Pi session, navigate to the folder, run the following command:

```
npm i --save johnny-five raspi-io
```

Then, run the program (be sure to use `sudo`!)

```
sudo node mqtt-lcd-button.js
```

Now, press the button and you should see the message pop up in the AdafruitIO feed dashboard:

And your LCD (remember, MQTT events are published to all, even the client that published them, if they are subscribed!):

While we're there, click **Actions**, then Add Data, and type `Hello from Adafruit!` in the data box, and hit **Create**. It should show up on your LCD:

And there you have it! You now have a bot that communicates with the internet via MQTT!

Project – social media notifier bot with IFTTT

Having lots of tabs open in your browser, and clicking on each one to see notifications, can be a nuisance. Luckily, we can easily build a project that pulls in notifications from several sources to create a bot that notifies us on an LCD. We'll also learn more about If This, Then That (IFTTT), and its plugin that will allow us to route events to Adafruit IO and thereby our Pi.

Getting started with IFTTT

IFTTT is a way to create graphical formulas (called applets) that consist of a trigger (such as a social media event) and an action (such as sending data to AdafruitIO). We're going to walk through linking AdafruitIO and our social media accounts to IFTTT, and creating Applets to send social media notification data to AdafruitIO.

First, sign in or create an account at `https://ifttt.com/`, and we'll start linking our accounts.

Linking IFTTT to Adafruit

To link your AdafruitIO account, click your username in the upper-right corner, select **Services**, then select the **All Services** link at the bottom of the page. Then, type `Adafruit` into the search bar, and click the black box marked Adafruit, which should bring you here:

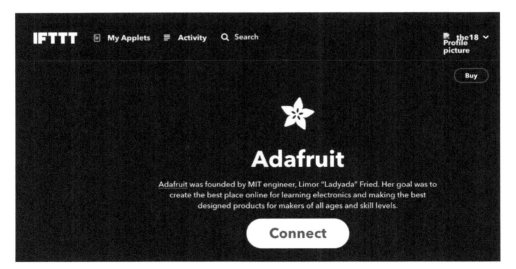

After that, click **Connect** and enter your AdafruitIO credentials. You'll want to do this for any social media networks that you'd like notifications from as well. I added `Twitter` and `Twitch.tv` accounts.

Now that we've linked our accounts, we should take a step back to Adafruit.IO and add feeds for our social media accounts so we can select them when we create our IFTTT applets.

Setting up feeds for your social media MQTT messages in AdafruitIO

In your AdafruitIO dashboard, you should create a feed (this corresponds to an MQTT topic) for each social media service you want your bot to notify you on. Having clear and granular topics is important when building large MQTT systems with several bots listening for events.

I created a group for these feeds by selecting **Actions** on the Feeds page of the AdafruitIO dashboard and clicking **Create a New Group**:

I named mine as `Social-Media-Bot`. Once you've finished making the group, click the name of the group to be taken to that group's dashboard, which should have no feeds in it yet. Click **Actions** and select **Create a New Feed** from the dropdown. Then, enter the name of the social media service you will be using as the name of the feed. Repeat for any social media services you'd like to use.

 The AdafruitIO Feed Group also serves as a handy namespacing tool. The twitter feed in the **Social-Media-Bot group** becomes **Social-Media-Bot.twitter**. This is extremely useful if you end up with multiple projects with Twitter data feeds.

Now that we have our social media accounts linked to IFTTT, and AdafruitIO feeds ready to receive data, let's create some IFTTT Applets to collect the social media notifications.

Creating our IFTTT Applets

On the IFTTT home page, you create a new applet by selecting your username in the top-right corner and selecting **New Applet**:

You'll be presented with the IFTTT Applet formula:

Click the **+this** link to be taken to a list of services you can use to trigger your IFTTT applet. Fill out the name of the social media service you wish to use in the search bar, and select it as it pops up.

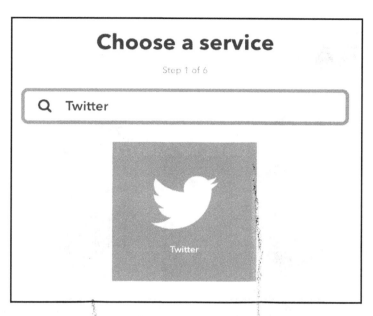

You'll then see a list of possible triggers from that service. Select which one you'd like to be notified for (I started with someone following my twitter account). Once you click your trigger, you'll be taken back to the formula page, and the **+this** will be replaced with the logo for the social media site you use as a trigger. Then, it's time to create our action by clicking **+that**.

You'll be taken to a similar page to select a service for your action. Search for `AdafruitIO` and select it. You'll be asked to fill out some information about which AdafruitIO feed you'd like to send to, and the message you'd like to send:

Select the name of the feed that matches the social media site that triggers this applet. Under **data to save**, you can enter a message that we can display on the LCD. You can also click the **Add ingredient** button to add information from the social media event itself:

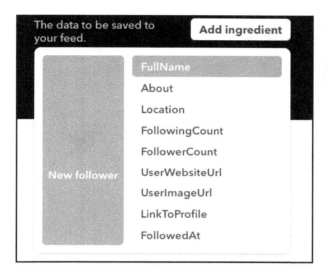

I selected **FullName** and ended up with the message `{{FullName}} followed you on Twitter!`

Repeat this for both the other triggers from the first social media site and the other social media sites and their triggers.

Now that our IFTTT Applets are sending data to AdafruitIO, we can start wiring and coding our bot.

Wiring up our project

All you need for this project is the LCD; if you wired up the button for the last project, you can remove it:

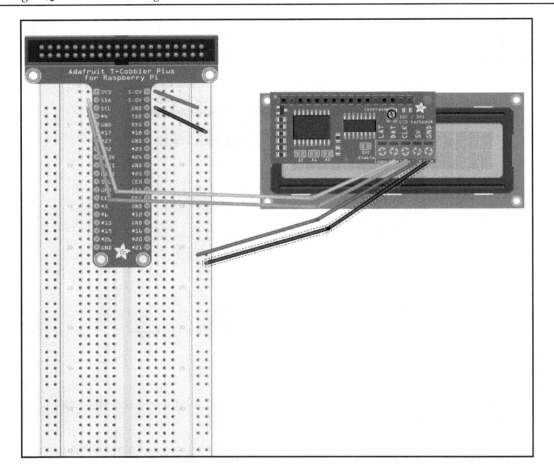

Coding our social media notifications to show on the LCD

Create a file named `social-media-bot.js` in your `project` folder, and copy the contents of `mqtt-button-lcd` into it. We're going to modify this file to create our social media bot.

First, remove all references to the button variable and `Button` object, since we won't be using them.

Next. we'll need to get the names of the feeds (topics) that we'll be subscribing to. To do this, click each feed in the AdafruitIO Feeds page that you want to use for this bot. Then in the right column of the page, click **Feed Information** and copy the **MQTT by key** field:

```
MQTT      nodebotanist/feeds/social-media-bot.twitter
by
Key
```

Then, in the `client.on('connect')` handler, we're going to subscribe to multiple topics, and not use the callback feature. Then, we'll add a `client.on('message')` handler to display the message from IFTTT on the LCD display:

```
client.on('connect', () => {
  console.log('Connected to AdafruitIO')
  client.subscribe('nodebotanist/feeds/social-media-bot.twitter')
  client.subscribe('nodebotanist/feeds/social-media-bot.twitch')
  client.on('message', (topic, message) => {
    LCD.clear()
    LCD.home()
    LCD.autoscroll()
    LCD.print(message)
  })
})
```

Now that we've subscribed to our new feeds and set it to print to the LCD, we can run it!

Running your social media bot

Run the following command to see the output:

```
sudo node social-media-bot.js
```

Soon, you should see a message pop up on your LCD when you get a social media notification.

Summary

In this chapter, we discussed several ways that internet-connected devices can talk to each other. We dove into the MQTT protocol and discussed how its PubSub interface and abstractions made it a great choice for projects on our Pi. Then, we built a bot that communicated with the outside world using a button and let the outside world communicate with it via the AdafruitIO dashboard and IFTTT. Finally, we built a social media notification bot using AdafruitIO integration into IFTTT.

Questions

1. What are two ways IoT devices interface with the internet that aren't specific to IoT?
2. What does MQTT stand for?
3. What is an MQTT client capable of?
4. What is an MQTT broker capable of?
5. Why does our Pi get the messages it publishes to AdafruitIO?

Further reading

- **A great tutorial on MQTT**: https://www.pubnub.com/blog/what-is-mqtt-use-cases/
- **The MQTT protocol official site**: http://mqtt.org/
- **AdafruitIO tutorials**: https://learn.adafruit.com/category/adafruit-io
- **The mqtt module page on npm**: https://www.npmjs.com/package/mqtt

Building a NodeBots Swarm **11**

Our NodeBots can now sense and display information about the world around them, and gather information from the internet for display. Now, it's time to look at NodeBots talking to each other. We will also use this chapter to talk about where to go from here—JavaScript is available on so many different boards and devices, and in so many new and exciting ways!

The following topics will be covered in this chapter:

- Project – connecting multiple NodeBots
- Expanding your NodeBots knowledge
- Continuing on your NodeBots adventure

Technical requirements

For this chapter, you can optionally use a second Raspberry Pi with Wireless internet access. You can also just use the one you have and we'll pretend it is two different Pis. You'll also need your TSL2591 Light Sensor, your Pi Cobbler, and a few breadboard wires.

 The code for this chapter is available at https://github.com/ PacktPublishing/Hands-On-Robotics-with-JavaScript/tree/master/ Chapter11.

Project – connecting multiple NodeBots

In this project, we'll use an npm module that allows us to set up a Raspberry Pi broker on a Pi—if you have one Pi, we'll have it talk to itself as if it's two separate devices, and if you have a second Pi, we'll have them talk to each other.

Optional – setting up a second Raspberry Pi

If you are only using your original Pi, skip this section.

Use the instructions in `Chapter 1`, *Setting Up Your Development Environment*, to set up your second Pi. You do not need a cobbler or any other accessories for this project; if you use a second Pi, you just need a good power source and a microSD card set up as stated in `Chapter 1`, *Setting Up Your Development Environment*.

If you are using two Pis, the Pi without the cobbler is the broker Pi, and your original Pi with the cobbler is the client Pi.

If you are using two Pis, you will want to set the hostname of the broker Pi so that they don't collide. To do this, run the following command:

```
sudo raspi-config
```

Then select **Network Options**, followed by **Hostname** with the arrow keys. Set the hostname to something you'll remember (I used `nodebotanist-pi-broker.local`). Then, save and exit `raspi-config` and reboot the Pi.

When you want to start a session on the broker Pi, you'll now use your custom hostname; for example, if I want to SSH into my broker Pi, I run the following command:

```
ssh pi@nodebotanist-pi-broker.local
```

Setting up your project files and folders

You'll want to create two separate project folders for this project: `client` and `broker`. Create these folders and run the following command:

```
npm init -y
```

In the `client` folder, run the following command:

```
npm i --save mqtt dotenv
```

In the `broker` folder, run the following command:

```
npm i --save mosca
```

Then, in both folders, create an `index.js` and a `.env` file.

If you're using one Pi

Move both the `client` and `broker` folders to the Pi, navigate to the `client` folder in your Pi SSH session, and run the following commands:

```
sudo npm i -g forever
npm i --save johnny-five raspi-io
npm i
```

Then, navigate to the `broker` folder and run the following commands:

```
npm i
sudo apt-get install mongodb-server // installs mongodb
systemctl enable mongod.service // makes it so mongodb starts when the pi
does
```

Then, reboot the Pi:

```
sudo reboot
```

If you're using two Pis

Move the `client` folder to your original Pi, SSH into your original Pi, navigate to the `client` folder, and run the following commands:

```
npm i --save johnny-five raspi-io
npm i
```

Then, move the `broker` folder to the second Pi you set up, SSH into it, navigate to the `broker` folder, and run the following commands:

```
sudo apt-get install mongodb-server // installs mongodb
systemctl enable mongod.service // makes it so mongodb starts when the pi
does
npm i
```

Then, reboot the broker Pi:

```
sudo reboot
```

Now that we have our project dependencies in place and installed, it's time to wire up this project up.

Adding a light sensor to the Pi

If you are using one Pi, wire the light sensor to it. If you are using two Pis, wire the light sensor to the client Pi. The following diagram should match either your sole Pi or your client Pi:

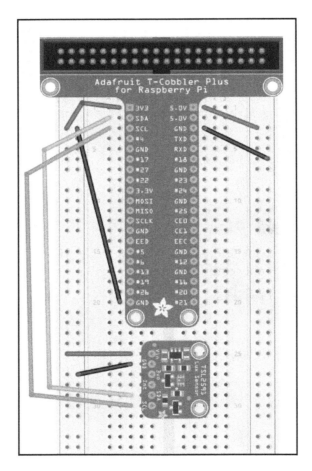

Now we can set up the MQTT broker on the Pi.

Creating an MQTT broker on the Pi

If you are using two Pis, carry out this entire section on the broker Pi. If you are using one Pi, do all of this on your one Pi.

We're going to use the Mosca library to set up an MQTT broker on our Pi. The `mosca` npm library makes it really easy to set up and start an MQTT broker. All we need is a running mongoDB instance (which we took care of in the last step, *Setting up light sensor*).

In the `broker` folder (either on your original or broker Pi), in the `index.js` file, we're going to set up `mosca`:

```
const mosca = require('mosca')

const mqttBroker = new mosca.Server({
  port: 1883,
  backend: {
    type: 'mongo',
    url: 'mongodb://localhost:27017/mqtt',
    pubsubCollection: 'MQTT-broker-NodeBots',
    mongo: {}
  }
})

server.on('ready', () => {
  console.log('MQTT broker ready!')
})

server.on('clientConnected', (client) => {
  console.log('Client connected to MQTT broker: ', client.id)
})

server.on('published', (packet, client) => {
  client = client || {id: 'N/A'} // Catches a weird edge case with mosca
  console.log(`Client: ${client.id}\nTopic: ${packet.topic}\nMessage:
${packet.payload.toString()}\n`)
})
```

We now have our MQTT broker ready to go! Time to program our client.

Programming the MQTT client – have the Pi Report Home

In your `client` folder (and your client Pi if you're using two Pis), open up the `index.js` file and write a script to gather light sensor data every time it changes with a threshold of `10` (to prevent too many MQTT messages):

```
const Raspi = require('raspi-io')
const five = require('johnny-five')
```

```
const request = require('request')

const board = new five.Board({
  io: new Raspi()
})

board.on('ready', () => {
    let light = new five.Light({
        controller: 'TSL2591',
        threshold: 10
    })

    light.on('change', () => {

    })
})
```

Then, before the `board.on('ready')` handler, construct your MQTT client connection, and add an `mqttClient.on()` handler that subscribes to the `light` topic:

```
const mqttClient = mqtt.connect(
  process.env.MQTT_BROKER_URL,
  {
    port: process.env.MQTT_BROKER_PORT
  }
)

mqttClient.on('connect', () => {
  mqttClient.subscribe('light')
})
```

Then, inside the `board.on('ready')` handler, we'll add the code that publishes light data to our MQTT broker.

```
light.on('change', function() {
  mqttClient.publish('light', this.value)
})
```

Now that we've coded our client, we need to set environment variables and get it running.

If you're using one Pi

In the `client` folder, create a `.env` file, and add the following:

```
MQTT_BROKER_URL:mqtt://localhost
MQTT_BROKER_PORT: 1883
```

Make sure to move the client and broker folders to your Pi one last time.

If you're using two Pis

On the client Pi, in the client folder, create a .env file with the following:

```
MQTT_BROKER_URL:mqtt://[broker pi hostname]
MQTT_BROKER_PORT: 1883
```

Replace [broker Pi hostname] with the hostname you created back in the *Setting up a Second Raspberry Pi* section.

Move the client folder to the client Pi one last time.

Now, it's time to get the code running!

Running our MQTT project

The instructions are slightly different for the one Pi and two Pi setups, but the end result should look the same.

If you're using one Pi

SSH into your Pi, navigate to the client folder, and run the following command:

```
su - pi -c "node forever start index.js"
```

This will cause our client to run in the background and re-start if necessary, so we can see the console.log() output from our broker. It will also ensure that our script is running as root, so that the Johnny-Five code will work properly.

Then, navigate to the broker folder, and run the following command:

```
sudo node index.js
```

You should start to see broker messages moving on the console as you change the light on the sensor; this is the MQTT client we set up in our client folder communicating with the MQTT broker on a different port.

If you're using two Pis

SSH into your client Pi, navigate to the client folder, and run the following command:

```
su - pi -c "node forever start index.js"
```

This will cause our client to run in the background and re-start if necessary, so we can see the console.log() output from our broker. It also ensures that our script is run as root, so that the Johnny-Five code will work properly.

Then, SSH into your broker Pi, navigate to the broker folder, and run the following command:

```
sudo node index.js
```

You should start to see broker messages moving through on the console as you change the light on the sensor; this is the two Pis speaking to each other using MQTT! Your client Pi is publishing messages to the broker, which console.log() them, but you could also connect with other clients that use this data!

You've now built the beginning of your first NodeBots swarm! Now it's time to take a peek at the wide world of NodeBots that falls outside the scope of this book.

Expanding your NodeBots knowledge

The NodeBots universe is huge and expanding everyday! In fact, we've only really started exploring the world of Johnny-Five.

Using Johnny-Five on other boards

The Raspberry Pi we used in this book is just one of over 40 boards supported by Johnny-Five. Just a peek at the platform support page of the Johnny-Five website makes this clear:

Johnny-Five has been tested with a variety of Arduino-compatible Boards. For non-Arduino based projects, platform-specific IO Plugins are available. IO Plugins allow Johnny-Five code to communicate with any hardware in whatever language that platform speaks!

The best news is that *the code you've written in this book can be transferred to most of the boards pictured here.* There are only two steps to porting your code: changing the pin numbers and making sure the board you are switching to offers the functionality you need.

Figuring out pin numbers

Changing pin numbers means you need to know what pins on your new board complete which tasks. For instance, if you were moving your I2C LCD from the Pi to the Arduino Uno, you'd need to know where the SDA and SCL pins are located on the Uno.

The best way to get this information is to search the internet for [board name] pinout, and searching for an image such as the following for the Uno:

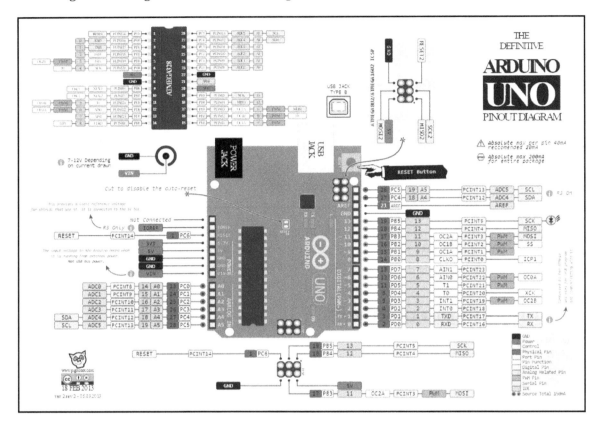

 A pinout of the ARDUINO Board and ATMega328PU (https://commons.wikimedia.org/wiki/File:Pinout_of_ARDUINO_Board_and_ATMega328PU.svg) by pighixxx is licensed under *Creative Commons Attribution-Share Alike 4.0 International* (https://en.wikipedia.org/wiki/Creative_Commons).

Then, you can match the pins up from there.

Checking the platform support page

Some boards support protocols and peripherals that others do not: the Arduino Uno has analog-in pins while the Pi does not, but the Pi has USB support while the Uno does not. Luckily, the Johnny-Five documentation, under *Platform Support*, tells you what is and isn't supported. The Uno is shown here as an example:

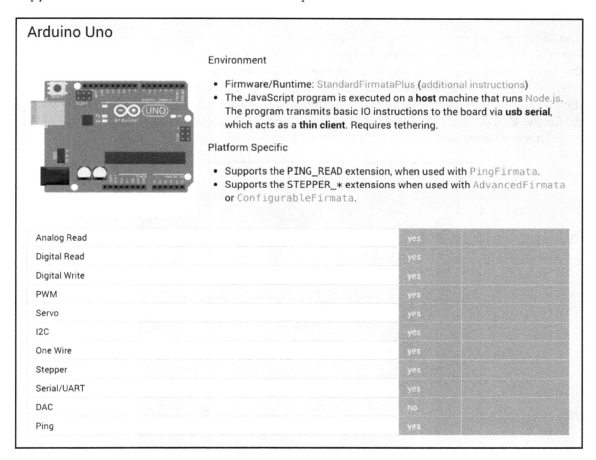

Now that you've got a glimpse at where you can continue your adventure within Johnny-Five, let's take a look at the even larger NodeBots world outside.

Other node robotics platforms

There's a wide world of NodeBots out there, and this list is by no means exhaustive. But let's take our first step into that wider world.

The Tessel 2

The Tessel project sought to create a Node.js native robotics project at a relatively low cost but with a great user experience, and they really have done a great job (disclaimer: I'm on the Tessel project team as a contributing member). In hardware terms, it's much like a Raspberry Pi—it runs Node.js on top of Linux; but not only is it a different form of Linux, the `tessel-cli` abstracts away much of the `ssh-ing` and Linux commands that we had to do for the Pi. The Tessel 2 is the current model, and it supports Johnny-Five right out of the box. To learn more, visit the Tessel project website at `https://tessel.io/`.

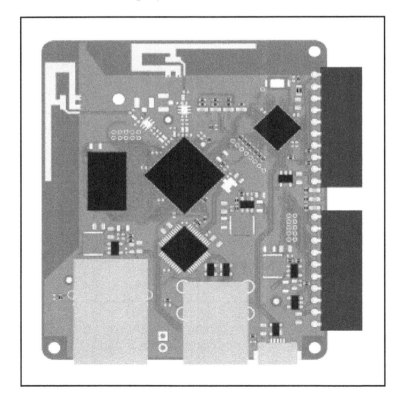

Image used with open-source license from https://github.com/tessel/project

The Espruino ecosystem

There are also bots outside the Johnny-Five ecosystem that run versions of JavaScript instead of full Node.js. One very popular set is the Espruino project boards. There's the Espruino main board, Espruino Wi-Fi, the Espruino Pico, Puck.js, Pixl.js, and an MDBT42Q breakout available at the time of writing. The Espruino project is headed up by Gordon Williams, you can find the boards at `https://www.adafruit.com/?q=Espruino,` while further information is available at `http://www.espruino.com/`.

The Espruino family holds a dear place in my heart, as it powered one of my first ever NodeBots, a light-up dress:

The author giving a panel talk at NodeConf US 2014 in her light-up dress shown in the above image

Programming graphically with Node-RED

Node-RED is a project that allows you to graphically program using blocks and write blocks using Node.js. Its graphical interface makes many tasks easier to grasp for younger soon-to-be programmers:

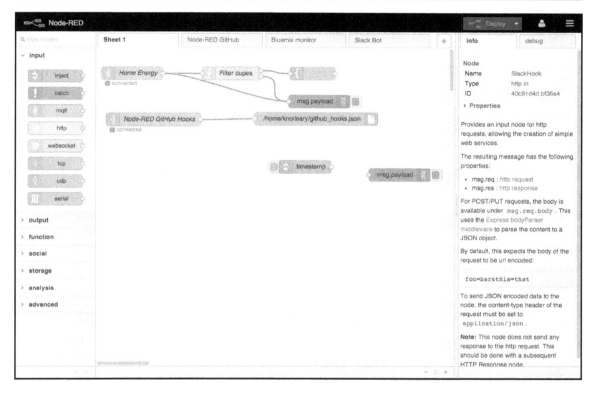

Image used with an open-source license from https://github.com/node-red/node-red

Again, this list just scratches the surface of the ever-burgeoning NodeBots world, and I encourage you to explore and find what works for you!

Continuing on your NodeBots adventure

Here are a few pieces of advice I've curated over the years to help you continue on your NodeBots journey; I hope they help you and I can't wait to see what you build!

Figuring out what to build

I tend to keep a note file open on my phone of cool things I'd like to have. Then, I go through that list and think: *can I build this?* I make sure not to consider whether I can buy it straight-up right away—sometimes it's more fun to build the thing you want instead of just buying it, and about 90% of the time, you end up being able to build a project that suits you instead of making do with a store-bought item that doesn't quite do everything you need.

Reaching out to the NodeBots community

Join us on Gitter at `https://gitter.im/rwaldron/johnny-five`. We love to help troubleshoot problems and answer questions! When you see someone doing NodeBots on Twitter, Reddit, and so on, make sure to reach out! Collaboration brings solutions and innovation, and it's a case of the more the merrier in the Johnny-Five and NodeBots community!

Where to go to learn more about Electronics

Here's a small selection of books that help teach electronics and related skills to those of you who aren't electrical engineers; I find them super helpful for when I have a hardware problem on my projects:

- *The Adafruit Guide to Excellent Soldering* (`https://learn.adafruit.com/adafruit-guide-excellent-soldering/tools`). If you don't know how to solder, or you've just been winging it, give this a read to make sure your soldering isn't what's stopping your project from working.
- *Practical Guide to Electronics for Inventors, Fourth Edition*, by Paul Scherz. This is a thick book and a dense read, but is a great reference work if you want to learn how electronic components work without doing differential equations.
- *Getting Started in Electronics*, by Forrest M Mims III. Do you need something a bit lighter than the Practical Guide? This book is a seminal work in hobbyist electronics, and this series of project notebooks teaches you how to use components to build fun and educational projects.

Summary

In this chapter, we completed our NodeBots journey together by learning how to get our NodeBots to talk to each other (or themselves). Then, we delved into the wider world of Johnny-Five and NodeBots. Finally, we looked at good reads to continue your quest for knowledge regarding electronics and related skills.

I love to see what my readers build—please feel free to reach me at @nodebotani.st or @nodebotanist on twitter to show me what you have come up with—even if it's a blinking LED and you're just proud to show it, I'm always happy to see it.

Thank you so much for reading. I appreciate my readers and wish you all the best!

Assessments

Chapter 1

1. We'll use the Linux operating system, via the Raspbian distribution, and leverage it to run our projects in Node.js.
2. **GPIO** stands for General-Purpose Input/Output.
3. Rick Waldron started the Johnny-Five project back in 2012 and wrote a program that used node-serialport to operate an Arduino Uno with Node.js.
4. We run `uname -m` command on the command line of our Pi ssh session to find out what ARM architecture the Raspberry Pi uses.
5. Changing the default Raspberry Pi password is important because default username and password isn't very secure, especially when your Pi is connected to the internet.
6. Node.js allows you to create even advanced robotics projects without having to deal with any low-level languages, also it has **Event-based systems** and **Garbage collection/automatic memory management**.
7. Node.js prides itself on creating small, bordering on tiny, packages, and has the excellent npm package manager (and others) to help manage those packages.

Chapter 2

1. The first `LED.strobe()` parameter defines the blinking speed of LED, which is 100 ms by default. If the argument is defined as 500 then the blink will go in 500 ms intervals.
2. The second line of argument use `require` function to pull in the `johnny-five` modules.
3. Johnny-Five LED objects are output only and therefore do not emit any events.
4. Raspberry Pi pin P1-29 translate to GPIO 5 in terms of GPIO#.
5. `LED.blink()` function is an alias to `LED.strobe()` function.
6. We begin with constructing a board object, and we pass a new instance of the `raspi-io` module in as its I/O. This is how we tell the Johnny-Five library we're running this code on a Raspberry Pi.

Chapter 3

1. PWM stands for Pulse-width modulation, it sets the percentage of time that a pin is HIGH and LOW. For an LED, it sets the effective brightness.
2. The Raspberry Pi 3 B+ has 2 PWM pins, but they operate on the same channel, effectively creating 1 PWM pin.
3. We need the GPIO expander because the RGB LED needs 3 PWM pins to fully function, and the Pi only has 1 on board.
4. 7 -- red, green, blue, white (red+green+blue), yellow (red+green), purple (red+blue), and cyan (green+blue).
5. The GPIO expander communicates with the Raspberry Pi using the I2C protocol.
6. The color module takes in strings representing color in various formats (#FFF, rgb(255,255,255), and translates them into a red, green, and blue channel that our Pi and LED understand.
7. The REPL helps with debugging by letting you see and manipulate the state the bot is in. It is powerful because most robotics platforms have a way of altering the state of the code while running.

Chapter 4

1. The events available to the Johnny-Five button object are `press`, `release`, and `hold`.
2. The Raspberry Pi cannot directly use analog input devices because all of its pins are digital.
3. We will use sensors with the Pi that add a digital interface to access the readings of the analog sensor itself.
4. There are no events for the Johnny-Five LED.RGB object because it strictly does output.

Chapter 5

1. An analog input sensor takes in data from its surroundings and converts it into a value that is represented by a voltage level sent either to an intermediate processor or the microcontroller directly.
2. Analog sensors cannot interface directly with the Pi because all of the Pi's GPIO pins are digital.

3. Two digital interfaces we can use to interface analog sensors with the Pi are I2C and SPI.

4. The two pins (besides ground and power) that an I2C device needs to operate are an SDA (data) pin and an SCL (clock) pin.

5. The sensor object can fire the `data` event, which means data has been collected, and the `change` event, which indicates that the data from the sensor has changed.

6. barcli is helpful in processing sensor data because instead of reading hundreds of lines of numbers, you can look at a bar graph and how it changes when you manipulate the sensor.

Chapter 6

1. A motor is an electrical device that converts electricity into rotational movement

2. The difference between a motor and a stepper motor is that motor can only be told what direction and speed to go, while a stepper motor can be told how many pre-defined increments to move, making it better for precision movements.

3. You should use external power for motors because otherwise, it will draw too much power from the Raspberry Pi, causing strange errors or even reboots while your project is running. Your motors may also run slowly or not respond to commands when powered directly from the Pi.

4. We need a Pi hat to control our motor because getting the motor to move backward requires extra components that the hat provides—it also makes it much easier to supply external power to the motors.

5. The benefits of the Motors object when controlling multiple motors is the ability to control all of your motors in the group with one command while retaining the ability to send commands to the individual motors.

Chapter 7

1. The difference between servos and motors is that you can tell a servo what angle to move to, while you can only tell a motor to go forwards or backward and at what speed.

2. The difference between regular and continuous servos is regular servos can go from 0-180 degrees and you can control what degree, while you can only tell a continuous servo which way to rotate and at what speed, but it has full 360-degree movement.

3. Servos require an external power source because otherwise, it will draw too much power from the Raspberry Pi, causing strange errors or even reboots while your project is running. Your motors may also run slowly or not respond to commands when powered directly from the Pi.

4. You would use servos over motors when you want your movements to go to a specific angle every time.

5. The benefits of the Johnny-Five Servos object is the ability to control all of your servos in the group with one command while retaining the ability to send commands to the individual servos.

Chapter 8

1. Animations are necessary for complex movements with multiple servos because timing is key for these movements, and moving all servos as fast as they can go with no easing makes complex movements nigh-impossible.

2. A keyframe is an array of information about the locations of the servos in an animation at an arbitrary point in time, defined by the animation segment.

3. A cue point is a point in the animation that lines up with a keyframe. When these are combined with duration in the animation segment, you will get a time for each keyframe.

4. The three pieces of an animation segment are the keyframes, the cue points, and the duration.

5. Easing manipulates our animation keyframes and segments by changing the acceleration rate of the servos as they travel from one keyframe to another.

6. The method of the animation object that stops the current segment and clears the animation queue is `Animation.stop()`.

7. Calling `Animation.speed(.25)` slows the currently running animation to 1/4 of its original speed.

Chapter 9

1. The Pi is well-suited for projects that require remote data because of its onboard ability to connect to the internet via either WiFi or Ethernet.

2. The considerations that need to be taken when making regular web requests from the Pi are:
   - The size of the payload
   - How much CPU parsing the payload will take

- Whether the data being accessed was meant to be accessed
- WiFi requests take a lot of power for the Pi and can cause problems if the Pi doesn't have a proper power supply

3. We can chain the LCD object calls, such as `LCD.clear().home()` because the Johnny-Five Objects always return the instance of the Object the method was working on, so another method can be called on it.

4. We use an I2C backpack with our LCD to cut the number of needed wires and pins from 8 to 2, and eliminate the need to hook up our own potentiometer to adjust LCD contrast.

5. We would need more components to use the LCD without the backpack—we would need a potentiometer to control the contrast of the LCD.

6. `LCD.on()` does not turn on the entire LCD, instead, it turns on the LCD backlight.

Chapter 10

1. Two ways IoT devices interface with the internet that aren't specific to IoT are HTTP/S and WebSockets.

2. MQTT stands for Message Queueing Telemetry Transport.

3. An MQTT client is capable of connecting to a broker, subscribing to topics, process incoming messages to topics they are subscribed to, and publishing to topics.

4. An MQTT broker is capable of letting MQTT clients connect, making sure published messages go to all clients that are subscribed to the message's topic, and publishing messages to any topic.

5. The Pi gets a copy of the messages we send to AdafruitIO because it is subscribed to the topic it is publishing to, and every device subscribed to the topic is sent the message by the broker.

Other Books You May Enjoy

If you enjoyed this book, you may be interested in these other books by Packt:

Learning Robotics using Python - Second Edition
Lentin Joseph

ISBN: 978-1-78862-331-5

- Design a differential robot from scratch
- Model a differential robot using ROS and URDF
- Simulate a differential robot using ROS and Gazebo
- Design robot hardware electronics
- Interface robot actuators with embedded boards
- Explore the interfacing of different 3D depth cameras in ROS
- Implement autonomous navigation in ChefBot
- Create a GUI for robot control

ROS Robotics Projects
Lentin Joseph

ISBN: 978-1-78355-471-3

- Create your own self-driving car using ROS
- Build an intelligent robotic application using deep learning and ROS
- Master 3D object recognition
- Control a robot using virtual reality and ROS
- Build your own AI chatter-bot using ROS
- Get to know all about the autonomous navigation of robots using ROS
- Understand face detection and tracking using ROS
- Get to grips with teleoperating robots using hand gestures
- Build ROS-based applications using Matlab and Android
- Build interactive applications using TurtleBot

Leave a review - let other readers know what you think

Please share your thoughts on this book with others by leaving a review on the site that you bought it from. If you purchased the book from Amazon, please leave us an honest review on this book's Amazon page. This is vital so that other potential readers can see and use your unbiased opinion to make purchasing decisions, we can understand what our customers think about our products, and our authors can see your feedback on the title that they have worked with Packt to create. It will only take a few minutes of your time, but is valuable to other potential customers, our authors, and Packt. Thank you!

Index

www.ingramcontent.com/pod-product-compliance
Lightning Source LLC
Chambersburg PA
CBHW080526060326
40690CB00022B/5036